PROTECTION OF
INDUSTRIAL POWER
SYSTEMS

OTHER TITLES IN THE SERIES

NOTICE TO READERS

PROTECTION OF INDUSTRIAL POWER SYSTEMS

T. DAVIES
British Steel Corporation, Cleveland, UK

PERGAMON PRESS
OXFORD · NEW YORK · TORONTO · SYDNEY · PARIS · FRANKFURT

U.K.	Pergamon Press Ltd., Headington Hill Hall, Oxford OX3 0BW, England
U.S.A.	Pergamon Press Inc., Maxwell House, Fairview Park, Elmsford, New York 10523, U.S.A.
CANADA	Pergamon Press Canada Ltd., Suite 104, 150 Consumers Road, Willowdale, Ontario M2J 1P9, Canada
AUSTRALIA	Pergamon Press (Aust.) Pty. Ltd., P.O. Box 544, Potts Point, N.S.W. 2011, Australia
FRANCE	Pergamon Press SARL, 24 rue des Ecoles, 75240 Paris, Cedex 05, France
FEDERAL REPUBLIC OF GERMANY	Pergamon Press GmbH, Hammerweg 6, D-6242 Kronberg-Taunus, Federal Republic of Germany

First edition 1984

Library of Congress Cataloging in Publication Data
Davies, T.
Protection of industrial power systems.
(The Pergamon materials engineering practice series)
Includes index.
1. Protective relays. 2. Electric power systems—
Protection. I. Title. II. Series
TK2861.D34   1983   621.319′24   83-4137

British Library Cataloguing in Publication Data
Davies, T.
Protection of industrial power systems.—
(Pergamon materials engineering practice series).
1. Electric power distribution 2. Industrial safety.
I. Title
621.319′24   TK3091
ISBN 0-08-029322-0 (hard cover)
ISBN 0-08-029321-2 (flexi cover)

Printed in Great Britain by A. Wheaton & Co. Ltd., Exeter

Preface

The protection which is installed on an industrial power system is likely to be subjected to more difficult conditions than the protection on any other kind of power system. The fault level may be variable and in some cases very limited; the supply arrangements can be altered by switching in or out interconnections between substations; the starting current of large motors may represent a significant proportion of the load current and so on.

In addition industrial power systems are often changed to accommodate additional plant with equipment installed by different contractors and although protection is provided the overall co-ordination of the scheme may be unsatisfactory.

This book is intended to help the Works Electrical Engineer and the Contractors' Engineer to achieve an understanding of the subject of protection.

The whole area of industrial power system protection is covered starting with the many simple devices which are used. These are usually based on the discrimination by time principle and a number are described. In addition the capabilities of the modern cartridge fuse are examined.

The more conventional types of relays which are used have more accurate operating characteristics. These are achieved by particular application of the basic relay elements so as to interpret the power system parameters.

The link between the power system and the relay is the current transformer and all aspects of its construction, design and operation are discussed in order that this important component can be fully understood.

Although it is not possible to divorce mathematics from fault calculations the method is simplified to the extent that little more than a knowledge of Ohm's law is required to tackle the three-phase and earth-fault calculations for any system.

The most widely used relay in any industrial power system is the inverse definite–minimum time overcurrent relay. Its application is critically examined so that optimum plug and time settings can be determined to provide a fully discriminative scheme.

The theory of the Merz–Price protection system from which most discrimination by comparison schemes are derived, the limitations which necessitate the modification to the Merz–Price system and the development of the high-impedance relay are discussed in detail. The application of the high-impedance relay to provide stability during through-faults and sensitivity to internal faults is demonstrated.

The application of relays to busbar protection schemes and all aspects of the protection of transformers, feeders and electrical machines are dealt with and the control circuits which are associated with protection.

There is a comprehensive description of the tests required following installation and those that are required during a maintenance programme. Finally the care of protection relays is covered.

In writing this book the words and experience of many people have been used, a lot of information and advice has been gratefully accepted and a great deal of help has been needed. To my friends and colleagues my sincerest thanks will, I hope, reduce my indebtedness to them.

It is, however, necessary to acknowledge particularly the work done by Mr. K. Preston and Mr. J. R. Barratt. Jim for his unfailing encouragement and advice and Ken for the technical back-up and editing at all stages.

I would also like to thank the British Steel Corporation.

Yarm, 1983

Contents

1 SIMPLE PROTECTION DEVICES   1

Direct-acting trip   2
Thermal trip   2
Time-limit fuses   4
Thermal overload devices   6
Oil dashpots   6
Fuses   8

2 RELAYS   15

Induction relays   15
Typical applications   18
Attracted-armature relays   23
Typical applications   25
Moving-coil relays   26
Thermal relays   29
Measurement   30
Timing relays   31
Design   32
Static relays   35

3 CURRENT AND VOLTAGE TRANSFORMERS FOR
PROTECTION   40

Current transformers   40
Construction   40
Design   41
Burden   42
Operation   43
Open-circuited current transformer   45
Short-time factor   46
Accuracy limit factor   46
Specification of current transformers   47
Rated secondary current   48
Secondary winding impedance   49

Primary windings   49
Application   49
Effect of CT magnetising current on relay setting   51
Quadrature or air-gap current transformers   51
Summation current transformer   51
Voltage transformers   52
Accuracy   53
Protection   53
Residual connection   53
Capacitor voltage transformers   55

4 FAULT CALCULATIONS   56

Impedance   56
Fault level   56
Generators   62
Cables   67
Source impedance   69
Motors   71
Practical example   71
Current distribution   77
Earth faults   78

5 TIME-GRADED OVERCURRENT PROTECTION   81

Settings   85
Time-multiplier setting   86
Application   87
Discrimination with fuses   90
Typical calculation   92
Earth-fault protection   98
Very inverse characteristic   100
Extremely inverse characteristic   101
High multiples of setting   101

6 UNIT PROTECTION   103

Relays   108
Application   109
CT knee-point voltage   112
Overall setting   113
The residual connection   113
Busbar protection   115

Phase and earth-fault schemes 120
Settings 122
Current setting 122
Duplicate busbar protection 124
Mesh-connected substation 124
Biased differential protection 125

7 TRANSFORMER PROTECTION 129

Types of fault 129
Differential protection 130
Magnetising inrush 131
Tap-changing 132
Biased systems 134
High-speed biased systems 134
Earth-fault protection 136
Protection of an earthed-star winding 139
Protection of the delta winding 139
Combination of overall and earth-fault schemes 140
Operation levels required for restricted earth-fault
   schemes 140
Standby earth-fault protection 140
Buchholz protection 140
Overcurrent protection 143
Instantaneous high-set overcurrent protection 143
Overload protection 144
Protection of a typical industrial installation 145
Restricted earth-fault protection 147
Balanced earth-fault protection 148
Overcurrent protection 149
Standby earth-fault protection 151

8 FEEDER PROTECTION 153

Differential protection 153
Factors affecting the design 154
Practical systems 159
Auxiliary equipment 167
Impedance protection 168

9 MOTOR PROTECTION 173

Overload protection 175

Insulation failure   178
Settings   179
Differential protection   182
Loss of supply   183
Synchronous motors   184

10   GENERATOR PROTECTION   186

Insulation failure   186
Earthing by resistor   186
Earthing by transformer   187
Stator protection   187
Earth-fault protection   189
Rotor earth-fault protection   189
Unsatisfactory operating conditions   190
Overcurrent protection   192
Overload   194
Failure of prime mover   195
Loss of field   195
Overspeed   196
Overvoltage   197
Protection of generator/transformer units   197
Unit transformers   199

11   CONTROL CIRCUITS   200

Batteries   204
Power factor correction   204

12   TESTING   210

Works tests   210
Tests on site   211
Commissioning tests   212
    CT polarity check   212
    CT magnetising characteristic curves   217
    Relay characteristic check   220
    Insulation tests   225
    Tripping circuit check   226
Routine maintenance tests   226
Test equipment   226
Care of protection relays   228

Index   229

Chapter 1

Simple Protection Devices

The function of a protection scheme is to ensure the maximum continuity of supply. This is done by determining the location of a fault and disconnecting the minimum amount of equipment necessary to clear it. When a fault occurs a number of relays will detect it but only the relays directly associated with the faulty equipment are required to operate. This is achieved by discrimination.

There are three methods of discrimination:

(1) Time, e.g. by using Inverse Definite Minimum Time Relays (IDMT) or impedance relays.
(2) Comparison, e.g. by using Differential Feeder Protection or Earth-Fault Relays. These are known as Unit Protection.
(3) Magnitude, e.g. by using High-set Overcurrent Protection.

Relays which discriminate by time protect the equipment with which they are associated and also act as back-up protection for other relays. The major disadvantage is that there is a delay in the removal of the fault which increases damage to the faulty equipment and increases the possibility of damage to healthy equipment which is carrying the fault current.

Relays which discriminate by comparison protect only the equipment with which they are associated and are therefore known as unit protection. The basic principle of unit protection is that if the current entering and leaving the unit is the same then it is healthy, if there is a difference then the unit is faulty. As unit schemes are stable even when passing high through-fault currents they can have low settings, usually well below nominal full-load current, and fast operating times.

Relays which discriminate by magnitude can only be used where there is a large change in fault current as is the case between the primary and secondary windings of a transformer.

Discrimination by time is the basis for many simple protection devices. The time delay being in general inversely proportional to current level. These schemes are applied to medium voltage systems

1

as an integral part of the circuit-breaker where it would be uneconomical to provide protection relays.

DIRECT-ACTING TRIP

Possibly the simplest of all schemes is the direct-acting trip where the actual fault current is used by passing it through one or two turns to provide a magnetic field which actuates the circuit-breaker trip mechanism to clear the fault current. On the larger circuit-breakers, say 400 A and above, there is a certain amount of adjustment which can be made to the tripping level by the winding up of a spring which will increase the force required to actuate the tripping mechanism. The adjustment is typically over a range of 5 to 10 times the circuit-breaker rated current.

THERMAL TRIP

Lower levels of fault current from a little over the rated load current of the circuit-breaker up to the magnetic instantaneous tripping device setting the protection is by a thermal tripping device. This consists of a bimetallic strip which is deflected by the heat generated in the strip by the fault current and eventually trips the circuit-breaker. The arrangement provides a time delay which is inversely proportional to current.

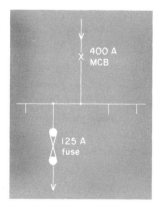

FIG. 1.1   MEDIUM VOLTAGE DISTRIBUTION SHOWING CIRCUIT-BREAKER AND FUSE

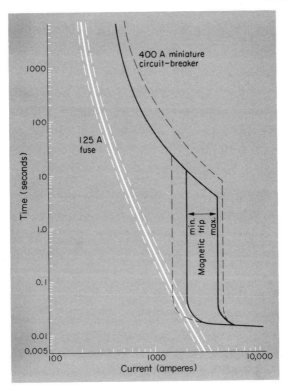

FIG. 1.2   TIME/CURRENT CHARACTERISTICS OF CIRCUIT-BREAKER AND FUSE

The usual application for this type of circuit-breaker is subcircuits or in situations where it is required to discriminate with fuses. Figure 1.1 shows a typical medium voltage distribution circuit and Fig. 1.2 the time/current characteristics of the circuit-breaker and the 125 A fuse. The full lines are the characteristic curves, the dotted lines the maximum tolerances. The characteristic curves suggest that the fuse will operate in a shorter time than the circuit-breaker at all current levels. However, to be certain of discrimination by time the magnetic trip setting must be increased to a 3000 A setting which allows discrimination to be achieved when the fuse time errors are positive and the circuit-breaker time errors are negative. Figure 1.3 shows this discrimination. These principles are used in devices known as miniature circuit-breakers (MCB) or moulded-case circuit-breakers (MCCB).

It will be noted that in protection discrimination diagrams the log-log graph is used extensively. It enables a wide range of times and current to be represented in a compact way.

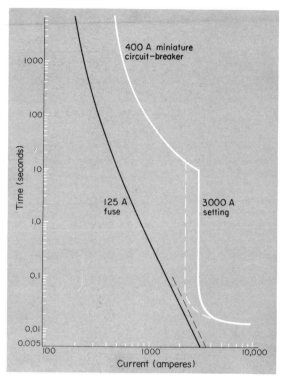

Fig. 1.3 Time/Current Characteristics of Circuit-breaker and Fuse Settings Arranged to Ensure Discrimination

TIME-LIMIT FUSES

An early form of inverse time protection was by the use of switchgear with a.c. tripping coils energised from current transformers. The time delay was provided by fuses which were connected so as to short-circuit the trip coils. Although this type of equipment would not be installed today there is still sufficient in commission to warrant an explanation of its operation.

Figure 1.4 shows the connections. When a fault occurs the current transformer secondary current is passed through the fuse and causes it to blow after a time which depends on the fault level and then causes tripping by energising the trip coil. Discrimination is achieved by progressively increasing the size of fuse as shown in Fig. 1.5.

Although time/current curves are published for the fuses it would be unwise to rely on calculated values of time as the proportion of current shunted by the fuse may be less than expected owing to the resistance of the connections and there is the likelihood of a change in

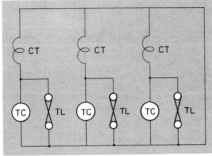

TC – Trip coil
TL – Time-limit fuse
CT – Current transformer

FIG. 1.4   TIME-LIMIT FUSE TRIPPING ARRANGEMENT

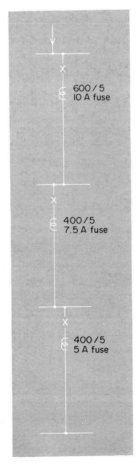

600 / 5
10 A fuse

400 / 5
7.5 A fuse

400 / 5
5 A fuse

FIG. 1.5   DISCRIMINATION USING TIME-LIMIT FUSES

fuse characteristic, particularly if it has been subjected to fault current which was cleared by another circuit-breaker.

The published curves are plotted in secondary current and therefore if a discrimination diagram was to be constructed the current transformer ratio should be taken into account and the diagram plotted for actual (primary) current.

Two other methods of protection are these days associated only with the tripping of contactors in motor circuits. These are thermal devices, where operation is via directly or indirectly heated bimetallic strip and magnetically operated overload devices where the time delay is by means of an oil dashpot. Formerly these devices were used to provide time discrimination in distribution circuit-breakers but are no longer used for this application.

THERMAL OVERLOAD DEVICES

These are used on motors up to 660 V and below, say, 50 kW. They are usually an integral part of the contactor and cause tripping by breaking the contractor holding-coil circuit. The heaters are connected one in each phase in the main circuit and carry the actual motor current and only for large motors would current transformers be used. The indirectly heated device usually has a heater winding wrapped around the bimetal whereas the bimetal itself is shaped so that it becomes the path for the current in the case of a directly heated device.

There is a comprehensive range of heaters from which a suitable one should be chosen to match the rating of the motor. The characteristic is an inverse time/current curve in the 10 s to 100 s region. The time setting is not adjustable but the current setting can usually be adjusted over a small range by a cam or screw.

In some cases there is, in addition to the normal overload arrangement, a single phasing detector. This operates the relay if there is a difference in deflection between the bimetallic strips in the three phases. This arrangement is not very sensitive and requires at least 80% of full-load current in one or two phases before tripping takes place.

OIL DASHPOTS

The oil dashpot is a means of producing a time delay and is usually used in conjunction with a solenoid-operated magnetic overload relay. In this type of relay there is a steel plunger with a piston at its lower end and the upper end inside a solenoid. The current setting is

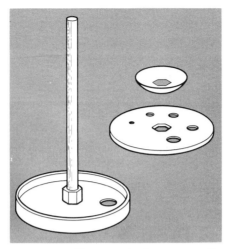

FIG. 1.6   OIL DASHPOT PISTON, WASHER AND RETAINING CLIP

determined by the initial position of the plunger in the solenoid and adjustment of this position can usually be made by screwing up or down the oil dashpot. The time delay is provided by the piston being moved through the oil in the dashpot. The actual time delay is dependent on the speed at which the oil can pass the piston and also on the viscosity of the oil. Modern silicone oils which have a low change in viscosity over a wide range of temperature make the device virtually independent of temperature. The speed at which oil passes the piston can be altered by having a washer which has a series of holes of different sizes and an arrangement to align one hole with a larger hole in the piston through which the oil must pass. The smaller the hole, the longer the time delay.

As can be seen from Fig. 1.6 there are six possible positions for the washer—five various sized holes and a position where there is no hole. If the latter position is selected the time delay is dependent on the oil passing the edges of the piston. The washer acts as a valve. When the plunger is being pulled into the solenoid the washer is pressed against the piston. When the plunger is resetting the washer lifts to allow an increased flow of oil so as to increase resetting speed. The operating force is proportional to the square of the magnetic flux density which in turn depends on the position of the plunger, i.e. the amount of steel inside the solenoid. When the force exerted on the plunger is greater than the total weight then it starts to move towards operation. Resetting is by gravity.

When used with machines up to, say, 37 kW the operating solenoid is connected directly into the line circuit whereas above 50 kW current transformers would be used.

FUSES

Simplicity of construction which permits consistency of manufacture and therefore accuracy of characteristic is the most important attribute of the modern cartridge fuse. It enables the performance to be predicted under all circumstances and can be applied to limit the energy which would flow to a fault and also allows discrimination with other fuses.

Along with all other parts of a power system a fuse is subjected to various levels of current but unlike other parts of the system it must behave differently under different conditions.

1. Load Current

The fuse must be capable of carrying the nominal load current of the circuit continuously and therefore should have a rating at least as high as the load current. The actual fusing current is 1.25 to 1.5 times the rated current and this level of current would cause fusing in 1 to 4 hours depending on fuse size. There is a tolerance on the figure which says the fuse link must fuse at 1.6× rating and not fuse at 1.2× rating. In the case of fuses protecting circuits which are capacitive, for example power factor correction installations or fluorescent lighting, then a fuse not less than 1.5× nominal load current should be used because of the high current inrush which occurs when the capacitors are energised.

2. Temporary Overload

For example, the starting of motors. It is generally accepted that if a fuse is subjected to a current for a time less than that shown on the characteristic time/current curve for that fuse then there will be no deterioration in fuse performance. Use is made of the fact in the selection of the fuse. Published data on a particular fuse generally includes a selection guide for direct-on-line starting of motors having a starting current of 7× motor full-load current for a time of 10 s and 3.5× full-load current for a time of 20 s. Examination of the time/current characteristic will show that the operating time for the recommended fuse at a current of 7 times the maximum of the range of current quoted is 10 s. Therefore, some reduction in fuse rating may be possible where the starting current is less than 7 times full-load current or where the starting time is less than 10 s. On the other hand, a higher starting current or longer starting time would necessitate a larger fuse size being selected.

In addition, if the motor is to be started frequently then a larger fuse must be chosen to keep the fuse within its allowable maximum temperature. It is difficult to apply a general rule to the allowance which must be made as heating and cooling time constants will vary. It must be stressed that it is the period between starts, i.e. the rate of starting, that is important.

3. Low-level Faults

This can be said to be the current levels covered by the published time/current curves. That is 1.5 times the fuse nominal rating up to the current required to operate the fuse in 0.01 s. It is in this area that protection relays would be required to discriminate with the fuse and also the area in which a fuse would be required to discriminate with a subcircuit fuse which is supplied at a reduced fault level owing to the length and area of the cable.

Discrimination between fuses and protection relays is dealt with in Chapter 5 whereas discrimination between two fuses would more usually be arranged at a current level corresponding to the cut-off region where the fault is cleared in a fraction of a cycle.

4. High-level Faults

By far the most important attribute of the modern cartridge fuse is its ability to limit the fault current and therefore the thermal and mechanical stresses in all equipment connected to a faulty circuit.

At high-fault levels sufficient energy is absorbed by the fuse in a fraction of a cycle to melt the fuse element and therefore the potentially high-fault current, known as the prospective current, is not achieved. This limitation of energy not only reduces damage in the faulty equipment but also prevents damage in the healthy parts of the circuit. Each fuse has an energy rating which is specified by manufacturers as an I^2t value. This value is not an entity but merely a means of comparison between fuses although it can be used to determine the energy dissipated in any part of the circuit for which the resistance is known. That is, by multiplying the I^2t value by the resistance. Energy = I^2tR joules. The term I^2t is not strictly correct as the current will be increasing during the whole of the time that the fuse is absorbing sufficient energy to operate. The actual energy required being that area under the curve as shown in Fig. 1.7.

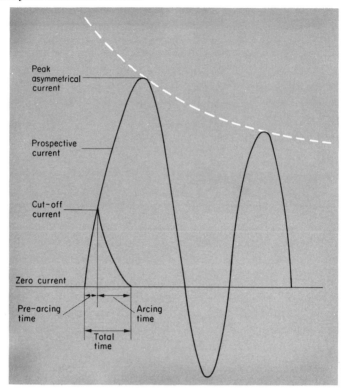

FIG. 1.7   FUSE OPERATION IN THE CUT-OFF REGION

Operation

There are two phases of operation of a fuse. The first period during which the energy to melt the fuse element is being absorbed—known as the pre-arcing time and the second period during which the arc is being extinguished—is known as the arcing time. The overall period is known as the total time.

The pre-arcing time is dependent on the current level whereas the arcing time depends on the system voltage. Both pre-arcing and arcing times are of the same order when the fuse operates at a high level of fault current, i.e. in a fraction of a cycle, but in the low-level fault region the arcing time is a very small proportion of the total time and is generally ignored.

Up to the point where sufficient energy is absorbed to operate the fuse there is no permanent change in its characteristic and it is this feature which is exploited to achieve discrimination. In the selection of a fuse the prime consideration must be given to the performance under normal healthy conditions. It must have a rating higher than the load current and must be capable of dealing with normal over-

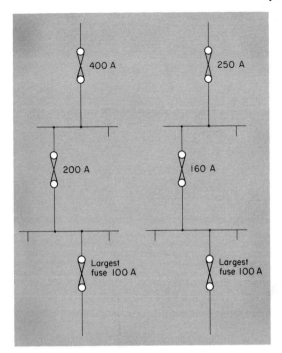

FIG. 1.8   DISCRIMINATION BETWEEN FUSES.
(a) USING 2 TO 1 RULE. (b) USING I^2t METHOD

loads, such as the starting of a motor, without being stressed. Once these criteria have been met consideration can be given to the question of discrimination.

Although a fuse is installed principally to interrupt short-circuit current and discrimination is regarded as secondary both requirements can be readily achieved. Discrimination between fuses is generally required in the high fault current region and if discrimination is achieved in this area then it is assured at all lower current levels.

One method of achieving discrimination is to double the size of fuse in each stage as shown in Fig. 1.8(a). There is no doubt that this method works. However, it is possible to effect some improvement. Manufacturers' I^2t values are sometimes presented in the manner shown in Fig. 1.9. At first sight it appears that the pre-arcing I^2t greatly exceeds the arcing I^2t but examination will show that the values are in fact comparable and that it is the logarithmic scale which makes it appear not to be so.

If the pre-arcing I^2t value is not exceeded then there will be no deterioration in the fuse characteristic. Therefore, if the total I^2t value of the smaller fuse does not exceed the pre-arcing I^2t value of the larger fuse then discrimination between the two fuses will be satisfactory. Figure 1.8(b) shows the reduction in fuse size which can be achieved by using the I^2t method.

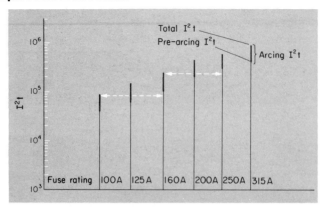

FIG. 1.9 FUSE I^2t CHARACTERISTICS

Protection of Motors

As previously mentioned selection of the fuse must not only take into account the motor full-load current but also the starting current, the starting time and the frequency of starting. It is not intended that the fuse should provide overload protection and indeed it will afford the motor very little protection, its main function being to protect the equipment between the fuses and the motor, i.e. the motor starter, cable and motor terminal box.

The current-breaking capacity of a motor starting contactor is generally arranged to be only slightly greater than the starting current of its motor. It is usually capable of making the maximum fault current which can flow in the circuit but not suitable for breaking current at this level. Hence any fault current in excess of the contactor breaking capability must be disconnected by the fuse.

Overload devices—thermal or magnetic—will operate to open the contactor at the lower levels of current but the fuse characteristic must be such that it takes over the clearing duty at a current less than the contactor breaking capability. Figure 1.10 is a typical time discrimination curve which shows that from 77 to 450 A tripping would be by the thermal overload relay which acts to open the contactor. Above 450 A the fault would be cleared by the fuse before operation of the overload relay therefore avoiding the possibility of the contactor being required to break current beyond its capability.

There has been some confusion with the term "motor circuit fuses". These are not fuses with special time/current characteristics but merely a means of extending the range of the standard cartridge fuse, which is the fuse normally used for motor circuit protection, by

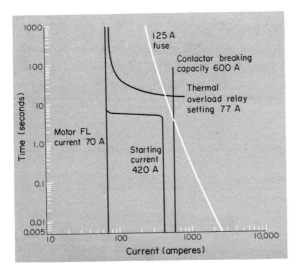

F<small>IG</small>. 1.10   C<small>OMPOSITE</small> T<small>IME</small>/C<small>URRENT</small> C<small>URVES</small>

assigning to it a continuous rating and a starting rating, e.g. 32M63. This means that whilst the time-current characteristic is the same as for a 63-A fuse, the physical dimensions are that of a 32-A fuse. It follows, therefore, that motor circuit fuses are manufactured with a value of continuous rating which corresponds to the value at which the fuse dimensions change, i.e. at 20, 32, 63, 100, 200, 315 and 400 A.

Cables

The IEE Wiring Regulations include protection of cables by fuses and an equation is given for the calculation of a relationship between fuse and cable characteristics. The equation is based on heating of the cable by fault current and is used to determine whether the disconnection time of the fuse is adequate to prevent conductor temperatures that would damage the cable insulation. The temperature rise of a cable is proportional to the energy loss in the cable as the loss of heat by conduction and convection is negligible in such a short time.

Semi-conductors

The silicon rectifier is now used extensively in industry both as a diode and a thyristor for power rectification, control and inverter circuits. The reason for this is that the losses in the device are extremely small and therefore considerable power can be handled by units having a very small physical size.

There is, however, a disadvantage in devices having a low mass and that is they have a very limited overload and overvoltage capacity and therefore the high current which can develop under fault conditions must be interrupted in a very short time. Unfortunately the rapid interruption of current produces very high voltages and therefore a fuse must also be used which has a low energy let through and also limits the overvoltage during interruption. Special fast-acting fuses have been developed for this application.

Chapter 2

Relays

When two protection devices are required to discriminate the chosen settings will depend on how closely the devices can be guaranteed to conform to their characteristic curves. Most of the devices covered in Chapter 1 have fairly generous tolerances in both operating levels and time and therefore if close discrimination is required then protection relays would have to be used.

A relay is a device which makes a measurement or receives a signal which causes it to operate and to effect the operation of other equipment.

A protection relay is a device which responds to abnormal conditions in an electrical power system to operate a circuit-breaker to disconnect the faulty section of the system with the minimum interruption of supply.

Many designs of relay elements have been produced but these are based on a few basic operating principles. The great majority of relays are in one of the following groups:

(1) Induction relays
(2) Attracted-armature relays
(3) Moving-coil relays
(4) Thermal relays
(5) Timing relays

INDUCTION RELAYS

Induction relays operate on the same principle as the induction motor. Torque is produced by subjecting a moving conductor to two alternating fields which are displaced in both space and time. The moving conductor is typically a metal disc which is pivotted so as to be free to rotate between the poles of two electromagnets. Torque is produced by the interaction of upper electromagnet flux and eddy currents induced in the disc by the lower electromagnet flux, and vice versa.

15

The torque produced is proportional to the product of upper and lower electromagnet fluxes and the sine of the angle between them.

$T \propto \Phi_a \Phi_b \sin A$.

This means that maximum torque is produced when the angle between the fluxes is 90° and as Φ_a and Φ_b are proportional to I_a and I_b

$T \propto I_a I_b \sin A$.

Consider the system shown in Fig. 2.1(a) and let I_a and I_b be in quadrature. This would be the condition if the upper coil, which is inductive, was supplied from system voltage and the lower coil with system current at unity power factor.

Φ_a and Φ_b the upper and lower electromagnet fluxes are phase with I_a and I_b respectively. Figure 2.1(b) shows the vector diagram and Fig. 2.1(c) shows the displacement in space of the relay pole faces. 1 and 5 are the outer poles of the upper electromagnet: 3 is the centre pole of

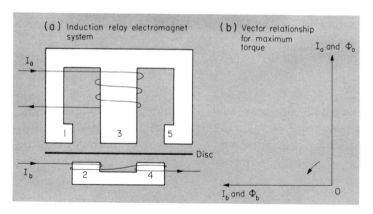

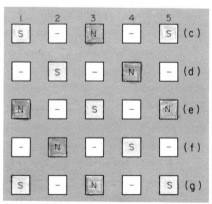

Sliding flux diagram

FIG. 2.1   SLIDING FLUX DIAGRAM

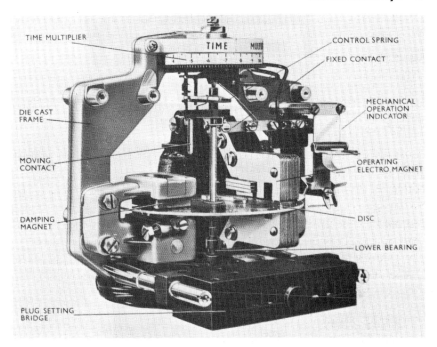

PLATE 2.1 INDUCTION DISC RELAY ELEMENT FOR SINGLE QUANTITY
MEASUREMENT (GEC Measurements)

this magnet and 2 and 4 are the poles of the lower electromagnet. At
the moment of time shown by the vector diagram, I_a is a maximum in
the positive direction and if pole 3 is assumed to be N then poles 1 and
5 are S. $I_b = 0$ and therefore poles 2 and 4 have no polarity.
One-quarter cycle later $I_a = 0$ and poles 1, 3 and 5 have no polarity. I_b
is a maximum in the negative direction and if pole 2 is assumed to be
S then pole 4 is N. This condition is shown in Fig. 2.1(d). Figs. 2.1(e),
2.1(f), and 2.1(g) show the conditions at ¼-cycle intervals. From this
diagram it can be seen that a sliding flux is produced which causes disc
movement from left to right. Reversal of polarity of either electromagnet will result in disc movement in the opposite direction.

Torque applied to a disc without control would, of course, continu-
ally accelerate the disc to a speed limited only by friction and
windage. Control is provided in two ways:

(1) By a permanent magnet whose field passes through the disc
    and produces a braking force proportional to disc speed. This
    controls the time characteristic of the relay.
(2) By a control spring which produces a torque proportional to
    disc angular displacement. This controls disc speed at low
    values of torque and determines the relay setting.

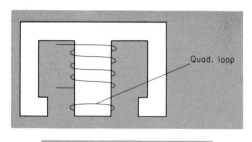

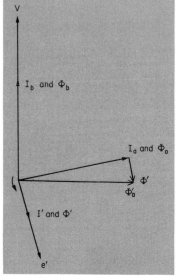

FIG. 2.2   THE EFFECT OF A QUAD. LOOP ON THE UPPER ELECTROMAGNET FLUX

From the above it can be seen that disc speed is dependent on torque, and as disc travel over a fixed distance is inversely proportional to time,

$$I_a I_b \sin A \propto \frac{1}{t}$$

which is an inverse time characteristic.

TYPICAL APPLICATIONS

(a) Wattmetric Relay

The upper coil is supplied by voltage and the lower coil by current. The voltage coil is very inductive and the voltage coil current, I_1, lags the voltage by about 80°. At unity power factor the phase angle between upper and lower coil fluxes would be about 80°. For a wattmetric relay it is necessary for maximum torque to be produced at unity power factor and therefore the fluxes must be 90° apart. To modify the upper coil flux a quadrature compensating loop, known as a quad loop, is used. This is merely a short-circuited turn of wire around the centre limb of the upper electromagnet. An e.m.f. is generated in the loop proportional to the rate of change of upper electromagnet flux. This e.m.f., which lags the upper electromagnet flux by 90°, produces a current I', which in turn produces a flux Φ', both flux and current are in phase with the e.m.f. The net effect is to produce a secondary flux, lagging the main flux by 90°, which modifies the upper electromagnet flux so that it lags the voltage by 90°. Where accurate measurement is required, e.g. kWh meters, the quad loop is made adjustable. The relay torque is therefore:

$$T \propto I_a I_b \sin (A + 90) \propto VI \cos A.$$

(b) kVAr Relay

For a wattmetric relay the correct phase angle is produced with, say, R/N voltage and R current. If R current was associated with Y/B voltage then the voltage phase shift is $-90°$ and the relay torque is

$$T \propto I_a I_b \sin (A + 90 - 90) \propto VI \sin A.$$

(c) Phase-angle-compensated Relay

From (a) and (b) it can be seen that relays having maximum response to any chosen phase angle can be produced. For example, Fig. 2.3 shows a relay with 45° compensated connections. Maximum

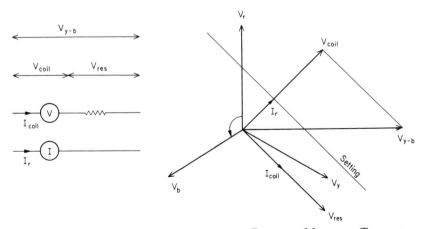

FIG. 2.3 INDUCTION RELAY COMPENSATED TO PRODUCE MAXIMUM TORQUE AT A SYSTEM PHASE ANGLE OF 45°

torque is produced when the current lags the voltage by 45° by associating R current with $Y-B$ voltage and connecting a resistor in series so that the voltage coil circuit current lags this voltage by 45°.

$T \propto I_a I_b \sin (A + 90 - 45) = VI \sin (A + 45)$.

Figure 2.3(b) shows the vector diagram for this connection.

(d) Overcurrent Relay

In an overcurrent relay a transformer connection is used. The upper electromagnet carries two windings, a primary which is fed from the current transformers and a secondary which feeds the lower electromagnet winding. As the secondary current is dependent on primary current and the phase angle between these is fixed, the relay torque is $T \propto I^2$.

(e) Over- or Undervoltage Relay

This is similar to (d) but the upper electromagnet winding is connected to the voltage supply. In the case of an undervoltage relay the contacts are arranged so that they make when the relay resets.

$T \propto V^2$.

In applications (d) and (e), that is where only one quantity is to be measured, then an electromagnet as shown in Fig. 2.4 may be used.

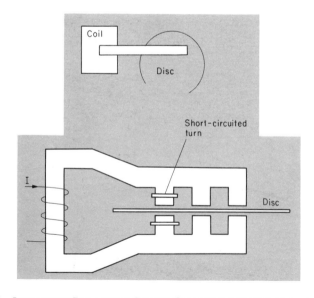

FIG. 2.4   INDUCTION RELAY FOR SINGLE QUANTITY MEASUREMENT

The short-circuited turn produces a phase displacement in the fluxes in adjacent poles causing movement of the disc. The torque is proportional to the square of the current.

Further application using this type of electromagnet are where the relay is required to respond to the sum or more usually the difference between two quantities, for example when used in a biased differential scheme where the vector sum is compared to the vector difference of two currents.

The full explanation of the use of this type of protection is given in Chapter 6 but the diagram shown in Fig. 2.5 gives a simple explanation. It is based on discrimination by comparison. If current flowing into the generator is the same as current flowing out then there is no fault and the relay should not operate.

If the currents are not the same then there is a fault and the relay should operate.

Coil A produces a torque in the disc in the direction to close the contact. The current in this coil is the vector sum of the input and output current—zero if there is no fault—whilst coil B produces a torque to open the contact. The current in this coil is the vector difference—maximum when there is no fault in the generator.

A further type of relay is the induction cup relay. The four-pole

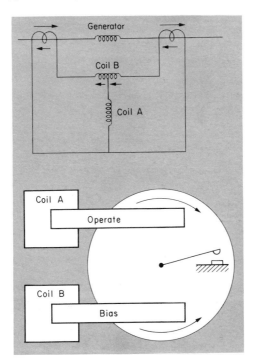

FIG. 2.5   PRINCIPLE OF BIASED DIFFERENTIAL PROTECTION

PLATE 2.2   INDUCTION CUP RELAY ELEMENT ASSEMBLY (GEC Measurements)

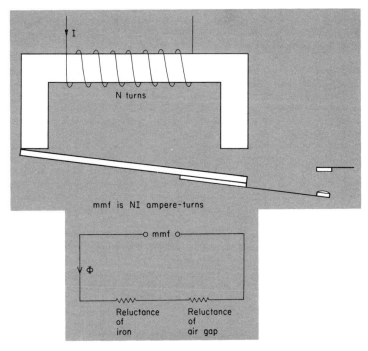

Fig. 2.6   Magnetic Circuit of an Attracted-armature Relay

electromagnet has an iron core and a copper cylinder which is free to rotate in the air gap between the pole faces and the core. This arrangement produces a high torque and is used mainly in high-speed protection schemes. As the air gap is small a high degree of accuracy is required in matching which makes it an expensive relay to manufacture.

The relay can be used as a simple product relay, e.g. $VI \cos A$, $VI \sin A$, etc., or in the eight-pole version as a polyphase device.

ATTRACTED-ARMATURE RELAYS

The attracted-armature relay comprises an iron-cored electromagnet which attracts an armature which is pivoted, hinged or otherwise supported to permit motion in the magnetic field.

The force exerted on the armature is given by the equation

$$\text{Force } F = \frac{B^2 s}{2\mu_0} \text{ newtons}$$

where B is the flux density in Wb/m^2, s is the effective area of the magnetic pole in m^2, and μ_0 is the permeability of free space $= 4\pi \times 10^{-7}$.

PLATE 2.3   ATTRACTED-ARMATURE CURRENT-MEASURING RELAY
(GEC Measurements)

The magnetic circuit can be represented in a similar manner to an electric circuit, Fig. 2.6, using magneto-motive force (m.m.f.) in ampere-turns applied to the reluctance of the iron and air gap in series—represented by resistance—which causes a flux Φ to flow in the circuit. The permeability of the iron is about 5000 times that of air which means that most of the m.m.f. will be used to magnetise the air gap. When the relay starts to operate, the length of the air gap, and therefore the reluctance, decreases which causes the flux, and the force, to increase. The effect of this in practical terms is that when the current in the coil reaches a value which produces sufficient force to move the armature—movement of the armature itself causes the flux and the operating forces to increase. So that once the armature moves it accelerates with increasing force until it is fully closed. This is the reason that contactors are very successful because once the contactor starts to move positive contact making is assured.

The snap action which is beneficial from the point of positive operation is sometimes a drawback in that the relay will not drop out until the flux density is reduced to below the pick-up value. As the magnetic circuit reluctance has been decreased by the closing of the armature a large reduction in ampere-turns is required to decrease the flux density to its original value, i.e. the relay has a low drop-off/ pick-up ratio. In some applications this can be inconvenient and in these instances the ratio can be improved by reducing the change in reluctance by not allowing the armature to close completely. In fact the ratio can be controlled by adjustment of the final air gap. An increase in drop-off/pick-up ratio reduces contact rating and operating speed.

In the simple case the moving contacts are carried by the armature but there are many cases where the armature is arranged to operate the contacts by means of a rod which pushes the contacts together (or apart if normally closed).

Control is generally by gravity assisted to a small extent by the contact spring pressure although in some cases spring control is used. Relays for use in a.c. circuits tend to vibrate no matter how large the operating quantity as the flux must pass through zero every half-cycle and during this period the armature tends to release. To eliminate this it is usual to split the electromagnetic pole face and surround one-half by copper loop. The current induced in this loop causes a phase delay in the flux passing through the loop compared with that passing through the other half of the pole and therefore the net flux is never zero. Alternatively vibration can be prevented by supplying the relay through a rectifier. In this case coil inductance maintains the flux during the idle portions of the cycle.

In d.c. operated relays residual flux is a problem and may prevent release of the armature. In order to reduce it to a low value the armature should not bed entirely on both poles of the electromagnet in the closed position but should always have a non-magnetic stop, to ensure that there is a small air gap.

In general attracted-armature relays are used

(a) as auxiliary repeat relays and for flag indicators. These are known as "all-or-nothing relays";

(b) as measuring relays where a drop-off/pick-up ratio of less than 90% can be tolerated.

TYPICAL APPLICATIONS

(a) All-or-nothing Relays

Tripping relays. These are multi-contact relays designed to be energised for a short time. The coil power is high resulting in an operating time of approx. 0.01 s. The relay can be self-resetting or of the latching type which are reset by hand or, with the addition of a second coil, electrically reset.

Auxiliary relays. These are for operation from the auxiliary d.c. supply and are used as repeat contactors to provide additional contacts and/or flag indicators with induction relays, moving coil relays or mechanical devices such as thermostats, buchholz relays, etc.

(b) Measuring Relays

The relay is suitable for all single quantity measurements, i.e. voltage, current, etc. Such relays usually have a range of adjustment by altering the number of effective turns on the coil in the case of current measuring relays; by changing the resistance in series with the coil in the case of voltage measuring relays or by adjustment of a spring so that the force required to pick up the relay can be changed.

MOVING-COIL RELAYS

The moving-coil relay consists of a light coil which when energised moves in a strong permanent magnet field. The coil can either be pivotted between bearings as in the usual moving-coil instrument (D'Arsonval movement) or suspended in the magnet field in the manner of the moving-coil loudspeaker (axial movement), Figs. 2.7

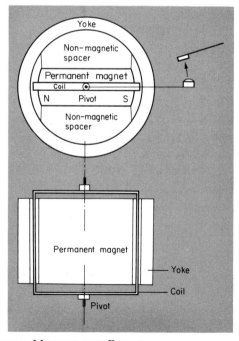

FIG. 2.7   D'ARSONVAL MOVING-COIL RELAY

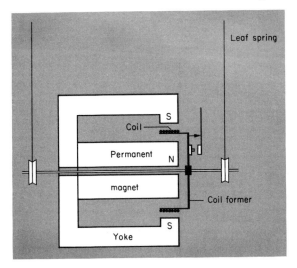

FIG. 2.8   AXIAL MOVING-COIL RELAY

and 2.8. In both cases the movement is very sensitive, that is, very little energy is required to produce operating force. It is for this reason, coupled with its ability to withstand high overloads, that it is almost invariably used in modern high-speed protection schemes.

The force produced is proportional to the product of the permanent magnet flux and the coil current. But, as the permanent magnet flux in any one relay is constant over the range of coil movement, the force is proportional to the coil current. The relay is polarised by the permanent magnet and must be used with a rectifier for all a.c. applications.

In the axial moving-coil relay the coil movement is essentially small whereas this need not be the case with the D'Arsonval relay. The latter, whilst it does have a short contact travel in its high-speed applications, can have a contact movement of up to 80°.

Movement damping is accomplished in both types by use of a metal coil former which acts as a shorted turn which will have current induced in it in such a direction as to oppose motion when the coil moves. In long travel relays the effect can also be used to introduce a time delay.

Control in both types of relay is by spring; leaf springs in the axial relay and a spiral spring in the D'Arsonval type. Current is conveyed to the coil and the moving contact carried by the coil by ligaments which in the D'Arsonval type are light spiral springs. The D'Arsonval movement is extremely sensitive, "galvanometer class" sensitivity is obtainable for special applications with a setting power as low as 20×10^{-6} watts. As the movement is proportional to current the

PLATE 2.4 MOVING-COIL RELAY ELEMENT (GEC Measurements)

contact differential ratio is nominally zero but on account of pivot friction, contact adhesion, etc., it is nominally 2%, i.e. drop-off/pick-up ratio = 98%.

In the long-travel relay it is usual to provide a calibrated scale along which the fixed contact can be set. In addition a 3-1 spring wind-up is allowable to widen the scale over the required operating range, e.g. an overvoltage relay could have a scale of, say, 100% to 150%. Two independently adjustable fixed contacts can be provided for use as low and high contacts with a side zero relay or forward and reverse with a centre zero relay.

The axial relay is less sensitive but is very robust. It has the advantage of having no bearings but on the other hand is affected by gravity if the relay case is not correctly aligned on the panel. In general moving-coil relays are used

(a) where a sensitive (low energy) relay is required,
(b) to provide a high drop-off/pick-up ratio,
(c) where the relay can be subjected to a continuous overload of many times its setting,
(d) in high-speed protection schemes.

The importance of a sensitive relay with a high overload capability can be appreciated when the conditions of operation of a protection scheme are considered.

A relay may be required to have a setting of, say, 10% normal current and yet be capable of carrying, say, 50 times normal current, which means that the relay must be capable of carrying 500 times setting current or 500^2 times setting power. With the moving-coil relay with a setting of 20 μW the power at maximum fault condition is only 5 W.

THERMAL RELAYS

These are relays in which the operating quantity generates heat in a resistance winding and so affects some temperature-sensitive component. Most protective relays of the thermal type are based upon the expansion of metal, a typical example being the use of bimetal material.

Bimetal is available in strips which are formed by welding two bars of different metals together throughout their length and then rolling out the composite bar to form a thin sheet. When a strip of this material is heated the difference in expansion rates of the two metals cause the strip to bend into a curve. The amount of motion of the end of the strip being magnified compared with the actual expansion of the individual metals. Relays can be constructed using straight pieces of bimetal or a longer strip may be coiled into a spiral thereby producing a large amount of motion in a compact space. The bimetal strip can be heated directly by passing current through it. In this case it is usually split longitudinally except at the extreme end so forming an elongated U. The two divided ends are clamped to a support and current is fed through the loop. This results in the bimetal becoming heated and causing motion of the tip through a proportionate angle.

MEASUREMENT

Single Quantity Measurement

This classification covers all simple relays such as those detecting current or voltage levels. The choice depends on the characteristic required.

Product Measurement

This subject has been partially discussed under induction relays where it was shown that the induction relay can readily be used to measure the product of two alternating quantities. The typical example of this is in power and directional types of relay.

It is interesting to note that other types of element can make a product measurement if the applied quantities are first mixed. For example, a beam relay is a natural amplitude comparator. If, however, two alternating signals A and B are first summated and then applied to the relay so that one coil is energised by the sum $A + B$, whilst the other coil is energised with $A - B$ then the relay becomes a phase comparator as the forces will only equal when there is a phase difference of 90° between A and B. Such an arrangement has been used as a directional relay.

On the other hand, if the two signals are summated as before and applied to the two windings of a power-type induction relay, then this combination will become a simple amplitude comparator because $(A + B)$ and $(A - B)$ have the same polarity only if A is greater than B.

These concepts are useful in designing new schemes with complicated response functions. When dealing with simple measurements it must be realised that some elements are fundamentally more suitable for amplitude or phase comparison than others, since notwithstanding the above algebraic conversions, errors must also be considered and these sometimes limit the range of application of an apparently suitable arrangement.

Comparators

Some means of comparing different quantities have already been mentioned, e.g. the induction relay and the balanced-beam type of relay. The latter is little used now but an equivalent element exists in the moving-coil relay with double windings. Such an arrangement operates in exactly the same way, to balance the two in feed currents.

A more effective means is the rectifier comparator bridge (Fig. 2.9). This comprises two full-wave bridge rectifiers each energised by separate transformer windings and connected on the d.c. side to circulate the current, i.e. positive of one rectifier connected to negative of the other. A moving-coil relay is connected across the bridge. When equal currents are fed into the two rectifiers their rectified components circulate and produce no current through the relay. When, however, one current exceeds the other the difference current flows in the relay. The relay can be made very sensitive since the maximum current to which it can be subjected is limited by the forward toe voltage of the rectifiers. Moreover, the rectifiers cannot be overstressed in the reverse sense since each provides a discharge path to limit the voltage for the other, the reverse voltage stress being limited to the forward voltage drop. By this technique and by special design of the feeding transformers the response characteristic of this arrangement can be very linear over a wide range.

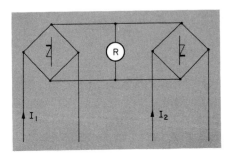

FIG. 2.9   RECTIFIER COMPARATOR BRIDGE

TIMING RELAYS

In some circumstances a time delay is required in conjunction with protection relays. These fall into three distinct groups.

(a) Short-time Relays

A short-time lag can be easily imposed using an attracted-armature type element, by fitting a solid copper cylinder to occupy a portion of the normal winding space. The "copper slug" may be placed at either end of the core but is most powerful when situated at the armature end. In this position it delays both operate and release functions of the relay by virtue of the eddy-currents induced in it which resist a change in the core flux. Time relays of the order of 50 ms in the operate sense and 200 ms for release are possible.

(b) Medium-value Accurate-time Delays

For this application a more elaborate mechanism is employed. The relay is powered by a solenoid or attracted armature element, either of which compress a spring. The other end of the spring drives a train of gears and an eddy-current brake system comprising a disc or drum rotating in a permanent-magnet field. The spring shaft also carries a contact arm which rotates as the gears run and ultimately makes contact at the end of its travel. A ratchet is usually fitted so that the relay can reset instantly when the coil is de-energised. This type of relay can give a maximum time delay in the range of 1.0 to 30 s and can be adjustable for any one value over a 10 to 1 range.

(c) Long-time Relays

Relays of this class are usually of the motor-operated type. The motor may be d.c. or a.c., either synchronous or induction, and will drive through gearing of such ratio that the operating time is achieved. The operating range extends from a few seconds up to hours, there being in principle no upper limit. When the gear ratio is high it is usual to incorporate a friction clutch in the drive chain, to avoid excess stress being built up should the motor continue to operate after the contact has completed full travel.

DESIGN

Many other designs of relay are possible and a great many other arrangements have been used and, providing that the necessary operating function is obtained, it only remains to say that the only other essential requirement is absolute reliability.

The protection relay, as distinct from a control relay, may remain inoperative for long periods but when operation is called for the response must be both immediate and accurate. For example, a busbar protection relay may operate under fault conditions perhaps only once in its normal span of life. If on this occasion should the relay be incapable of performing its function, owing to some deterioration which has taken place, then its provision has been in vain. Furthermore, the very fact that it has remained inactive for a long period is the condition which is liable to lead to the mechanism becoming stuck so as to be inoperative. Hence protective relays are designed with certain principles in mind:

(a) Simplicity

In so far as this is compatible with achieving the necessary measurements, simplicity is a most desirable characteristic. Any reduction in number of components reduces the possible causes of difficulty and simplicity in operation assists the maintenance staff and generally results in higher standard of maintenance.

(b) High Operating Force

Relays are designed with as high a working force as possible to minimise the effects of friction so that should it vary during the life of the relay the overall effect on the performance is negligible.

(c) High Contact Pressure

This is closely related to the working force but is also governed by the contact shape. To this end domed or cylindrical form contacts are used so that the contact-making area is small with the result that a given force corresponds to a high pressure.

(d) Contact Circuit Voltage

For general purposes contacts are made from silver which is excellent in its general characteristic. In bad atmospheres, however, it is liable to form surface layers of oxide or sulphide which are not a great detriment unless the layer is excessively thick. In general contact difficulties are encountered

(1) where there is a bad atmosphere,
(2) where the tripping voltage is low (30 V or less),
(3) with low torque relays, i.e. where contacts make on resetting or with thermal relays.

(e) Contact-making Action

Contacts should close together with a certain amount of wiping or scraping action in order to help in breaking down the surface films of oxide or other contaminants and should be designed so that they do not bounce apart or chatter after first closing. It is very difficult to ensure that the impact between the contact tips on making does not result in a rebound but the effect can be minimised by suitable design. Many complex arrangements have been evolved, but for normal purposes the main requirement is to ensure that the moving contact has a lower natural frequency than the fixed one. It is also important to ensure that the rest of the element and moving system does not generate excessive vibration which can be passed on to the contact. Any chattering from such a source might lead to excessive burning of the contact tips.

(f) Minimum Size of Wire

It is desirable that protective relay coils should not be wound with a wire which is thinner than 0.1 mm to guard against the risk of mechanical fracture. An even more serious problem is that of corrosion. It is most important that all the insulating materials with which the coil is wound should be absolutely neutral and incapable of releasing even small traces of substances with corrosive tendencies. Even when this is done coil corrosion can occur if the coil is allowed to assume a positive potential relative to earth. Should this be the case the wire is an anode and any small leakage current will deposit copper from the wire which in time will be corroded away by this electrolytic action. Even with moderate potentials and quite high insulation resistance to earth the wire can be completely severed by this action within quite a short time. Hence, it is desirable that all d.c. coils should be connected to the negative pole of the battery or maintained at a negative potential relative to earth by some other means.

(g) Enclosures

Even robust relays have to be regarded as precision measuring instruments and although they may work well when first produced they will not maintain this quality if exposed to accumulations of dust and other deposits from the atmosphere. Therefore the relay should be enclosed in a substantial and tight-fitting case which is made as dust-proof as possible by the fitting of gaskets although it should still be possible for the relay to breathe slightly. Under these circumstances the relay should remain in good condition for long periods.

STATIC RELAYS

Relays based on electronic techniques offer many advantages over the more usual electromechanical type. Apart from the obvious advantage of no moving parts the power requirements are low and therefore smaller current and voltage transformers can be used to provide the input. Additional benefits are improved accuracy and a wider range of characteristics.

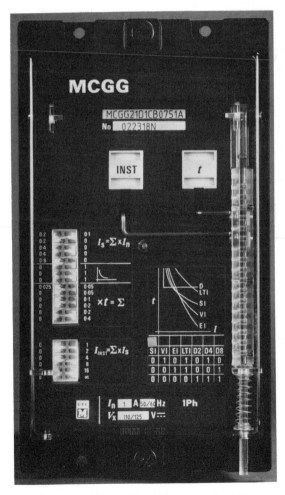

PLATE 2.5   MICROPROCESSOR-BASED OVERCURRENT RELAY
(GEC Measurements)

PLATE 2.6   TYPICAL INSTALLATION OF MICROPROCESSOR-BASED RELAYS
(GEC Measurements)

The invention of the transistor has allowed the development of static relays but difficulties were experienced because the high voltage substation proved to be a very hostile environment to the device. The close proximity of high voltage heavy current circuits produces conditions which could damage the transistor because of its low thermal mass or cause maloperation of the relay because of the electromagnetic or electrostatic interference.

A lot of research and development has taken place and commercial relays which meet very exacting standards have been produced. Electromechanical relays still represent the bulk of relays manufactured and it is unlikely that there will be a sweeping change-over to static relays particularly where the electromechanical relay is adequate. However, most of the future development in protection will be in static relays.

The large application potential of the digital integrated circuit has led to enormous expenditure on research and development which has resulted in microprocessors with spectacular computing capabilities at a low cost. It is fairly certain that microprocessors will ultimately dominate protection and control systems.

The utilisation of microprocessors in the field of protection means that the logic part of the relay can be replaced by a programme held in the microprocessor memory. This enables a relay function to be specified by software which widens the scope of the relay and allows a single relay to be provided with a number of characteristics.

Experience has been gained with microprocessors in high voltage substations over a number of years by using them for voltage control, automatic switching and reclosing and other control functions. Therefore difficulties which arise in this environment have been overcome.

An example of the versatility of the microprocessor is demonstrated in one of the first protection applications. This is an overcurrent relay which has a setting range of 10% to 200%, an extremely wide range made possible by the low power requirement of the relay, and a choice of five different characteristics. Figure 2.10 shows the block diagram of the relay.

The CT current is reduced to a more suitable level by an interposing current transformer in the relay. The current is rectified and passed through a resistance network which produces a voltage output which is proportional to current. The network provides the current setting control by switches mounted on the front of the relay and its output is fed into the analogue-digital converter which is part of the microprocessor.

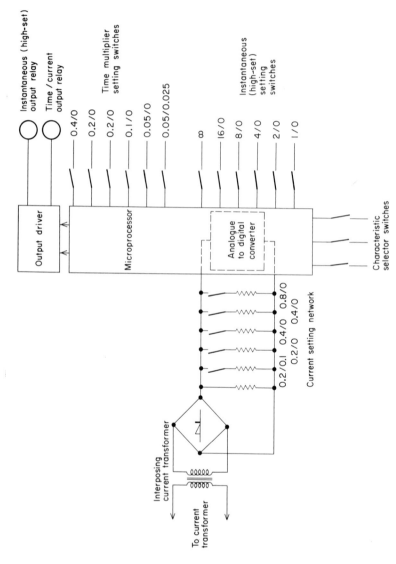

FIG. 2.10 SIMPLIFIED BLOCK DIAGRAM OF A MICROPROCESSOR-BASED OVER-CURRENT RELAY

Three other banks of switches are mounted on the front of the relay. The switches are connected to separate input ports on the microprocessor and control the time multiplier setting, the high-set relay setting and selection of the type of characteristic required, i.e. normal, very or extremely inverse, long time inverse or definite time.

Chapter 3

Current and Voltage Transformers for Protection

CURRENT TRANSFORMERS

The current transformer is well established but is generally regarded merely as a device which reproduces a primary current at a reduced level. A current transformer designed for measuring purposes operates over a range of current up to a specific rated value, which usually corresponds to the circuit normal rating, and has specified errors at that value. On the other hand, a protection current transformer is required to operate over a range of current many times the circuit rating and is frequently subjected to conditions greatly exceeding those which it would be subjected to as a measuring current transformer. Under such conditions the flux density corresponds to advanced saturation and the response during this and the initial transient period of short-circuit current is important.

It will be appreciated, therefore, that the method of specification of current transformers for measurement purposes is not necessarily satisfactory for those for protection. In addition an intimate knowledge of the operation current transformers is required in order to predict the performance of the protection. Current transformers have two important qualities:

1. They produce the primary current conditions at a much lower level so that the current can be carried by the small cross-sectional area cables associated with panel wiring and relays.
2. They provide an insulating barrier so that relays which are being used to protect high voltage equipment need only be insulated for a nominal 600 V.

CONSTRUCTION

Current transformers are usually designed so that the primary winding is the line conductor which is passed through an iron ring

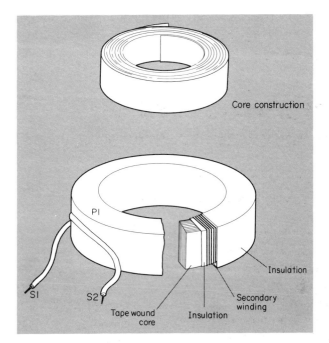

FIG. 3.1   CONSTRUCTION OF A CURRENT TRANSFORMER

which carries the secondary winding. They are mostly of this type and are known as bar-primary or ring-wound current transformers.

The construction of a typical ring-wound current transformer is shown in Fig. 3.1. Grain-oriented sheet-steel strip is wound to form a core and is covered with a layer of insulation. The secondary winding is wound over this and consists of the number of turns needed to produce the required ratio, of wire of sufficient cross-sectional area to carry rated current, followed by a further layer of insulation which covers the secondary winding. When installed the primary conductor, which acts as a single turn, is passed through the centre of the ring. The making of the core by stacked annular laminations has now been superseded by the wound-tape method.

DESIGN

Current transformers conform to the normal transformer e.m.f. equation where the average induced voltage is equal to the product of the number of turns and the rate of change of magnetic flux (Φ). The normal design criterion is to limit the flux to the value where saturation commences—known as the knee-point flux—and therefore it is the maximum value of the magnetising current which produces this flux. Magnetising current, and consequently flux, changes from zero to maximum in $\frac{1}{4}$ cycle and therefore the rate of change of flux is

$$\frac{\Phi - 0}{\frac{1}{4}} = 4\Phi \text{ webers/cycle}$$

or at a frequency of f cycles/s

$4\Phi f$ webers/s

giving an average induced voltage of

$V_{av} = 4\Phi fN$ where N is the number of turns

or in r.m.s. values, the knee-point voltage is

$V = 4.44\Phi fN$ as $V = 1.11V_{av}$,

also as flux Φ = flux density, B(tesla) $\times$ core area, s (m^2) the knee-point voltage is

$V = 4.44BsfN$.

The flux density of electrical sheet steel is about 1.5 tesla at knee-point which for a ring-type current transformer of known ratio makes the knee-point voltage fairly easy to estimate if the approximate dimensions of the core is known. For example, a CT ratio of 300/1 with a core area of 40 $\times$ 30 mm would have a knee-point flux of

$1.5 \times 40 \times 30 \times 10^{-6} = 0.0018$ weber

which on a 50-Hz system would produce a knee-point voltage of

$V = 4.44 \times 0.0018 \times 300 \times 50 = 120$ V.

BURDEN

The load of a current transformer is called the burden and can be expressed either as a VA load or as an impedance. In the former case the VA is taken to be at the CT nominal secondary current. For example, a 5-VA burden on a 1-A transformer would have an impedance of 5 ohms:

$$\frac{5 \text{ VA}}{1 \text{ A}} = 5 \text{ V}$$

$$\text{impedance} = \frac{5 \text{ V}}{1 \text{ A}} = 5 \text{ } \Omega$$

or on a 5-A current transformer:

$$\frac{5 \text{ VA}}{5 \text{ A}} = 1 \text{ V}$$

$$\text{impedance} = \frac{1 \text{ V}}{5 \text{ A}} = 0.2 \ \Omega.$$

All burdens are connected in series and the increase in impedance increases the burden on the current transformer. A current transformer is unloaded if the secondary winding is short-circuited as under this condition the VA burden is zero because the voltage is zero. The errors of transformation depend on the angle of the burden as well as its impedance.

OPERATION

A representation of a ring-type current transformer is shown in Fig. 3.2. R_2 is the secondary winding resistance, I_e the magnetising current and R_b and X_b are the burden resistance and reactance. The primary ampere-turns must equal the sum of the secondary ampere-turns and the magnetising ampere-turns.

$$N_1 I_1 = N_2 (I_2 + I_e).$$

In practice I_e is small compared to I_2 and is therefore ignored in all CT calculations with the exception of those concerned with ratio and phase angle error.

The magnetising current depends on the voltage V_2 which in turn depends on the product of the secondary current and the impedance of the burden plus the CT secondary winding resistance. That is, by Ohm's law,

$$V_2 = I_2 (R_2 + R_b + jX_b).$$

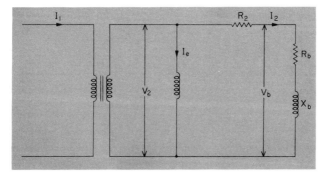

FIG. 3.2   EQUIVALENT CIRCUIT OF A RING-TYPE CURRENT TRANSFORMER

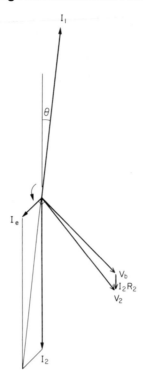

FIG. 3.3   VECTOR DIAGRAM OF A RING-TYPE CURRENT TRANSFORMER

If a vector diagram is drawn, Fig. 3.3, then the ratio error, which is the difference in magnitude of I_1 and I_2, and 0, the phase angle error, become apparent.

The magnetising current I_e lags V_2 by 90°. It can be seen that if the burden was wholly resistive then the ratio error would be a minimum and phase-angle error maximum, whereas if the burden was wholly reactive then the ratio error would be maximum and the phase-angle error minimum.

Note. The term $(R_2 + R_b + jX_b)$ is not a simple arithmetic sum as X_b is 90° out of phase with R_2 and R_b and so must be added by vectors. To denote this the prefix "*j*" is used which literally means "advance by 90°". The voltage I_2X_b is therefore 90° ahead of I_2R_2 and I_2R_b and $V_b = I_2(R_b + jX_b)$.

Figure 3.4 shows a magnetising characteristic for a 100/1 A current transformer. It has been previously stated that I_e is small compared to I_2 up to and beyond the knee-point of the characteristic. Hence the ratio and phase-angle errors will also be small. This means that the

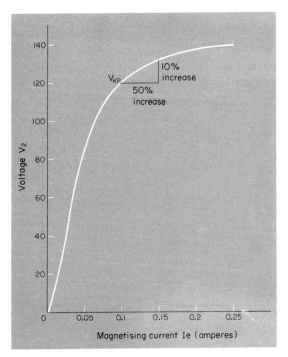

FIG. 3.4   CT MAGNETISING CHARACTERISTIC

primary/secondary current relationship will be maintained to this point,

i.e. where the product $I_2(R_2 + R_b + jX_b)$ is 120 V,

e.g. if $R_2 = 1\,\Omega$ and $R_b + jX_b = 7.5 + j0\,\Omega$ then linearity would be maintained up to a secondary current of

$$I_2 = \frac{V}{R_2 + R_b + jX_b} = \frac{120}{8.5} = 14.1\ \text{A or } 14.1 \times \text{CT rating.}$$

Alternatively, if linearity is required up to, say, 20 × CT rating then the total impedance should not exceed

$$R_2 + R_b + jX_b = \frac{120}{20} = 6\,\Omega.$$

OPEN-CIRCUITED CURRENT TRANSFORMER

If the impedance $R_b + jX_b$ is very high then the voltage calculated from $I_2(R_2 + R_b + jX_b)$ would be very large, well above knee-point value and I_e would become significantly large in the ampere-turn balance equation $N_1 I_1 = N_2(I_2 + I_e)$ and I_2 would be reduced. The limiting value is when the CT secondary winding is open-circuited and $I_2 = 0$. All the input ampere-turns will be used as magnetising ampere-turns and will drive the current transformer into saturation. As can be seen from Fig. 3.4 the greatly increased magnetising current will not cause much increase to the average voltage. However, the change in flux from zero to the knee-point value is not accomplished in $\frac{1}{4}$ cycle but in perhaps 1/100 of this time. Thus the rate of change of flux and, therefore, the induced voltage during this period would be about 100 times the knee-point voltage. Insulation can be damaged by this high short-duration voltage and overheating caused by the great increase of iron losses.

SHORT-TIME FACTOR

When a current transformer is used in a power system it may be subjected to fault current many times larger than its primary rating and, therefore, it must be able to withstand the effects of this current for the time for which it is likely to persist. The maximum current which it can carry without mechanical and thermal damage is expressed as a multiple of its rated current and is known as the "short-time factor". For example, a current transformer of ratio 200/5 which is capable of withstanding a current of, say, 13,000 A would have a short-time factor of 65. Such a short-time factor would always be associated with a period of duration of the current, for example 3 s. Smaller currents would be permissible for longer periods, the permissible time increasing as the square of the reduction of current. Larger currents, however, are not necessarily permissible for any period of time, since electromagnetic forces have also to be considered.

ACCURACY LIMIT FACTOR

When a current transformer is used to energise a protective relay it must maintain its characteristic ratio up to some multiple of its rated current. This multiple, which depends on the type and characteristics of the protection, may be 10, 20 or some even higher value and is known as the "Accuracy Limit Factor".

The small ratio error introduced by the magnetising current is often compensated for in the case of measuring current transformers by slightly modifying the ratio of primary to secondary turns from the nominal ratio. For example, a 100/1 current transformer might have one primary turn and 98 secondary turns so that the transformation ratio would appear to be 100 to 1.02 A, but when it is used to supply its rated burden the secondary current is reduced from the above value to 1 ampere by the magnetising losses.

Although the burden of a protective scheme is only a few VA at rated current, if the accuracy limit factor is high the output required from the current transformer may be considerable. On the other hand, it may be subjected to a very high burden. For example, in the case of overcurrent and earth-fault protection having elements of similar VA consumption at setting, if the overcurrent elements are set at 100% an earth-fault element set at 10% would have 100 times the impedance of the overcurrent elements. Although saturation of the relay elements modify this somewhat, it will be seen that the earth-fault element is a severe burden and the current transformer is liable to have considerable ratio error in this case. For this reason it is not very much use applying turns correction to current transformers used for protective purposes and it is generally simpler and more satisfactory to wind them with turns corresponding to the nominal ratio.

SPECIFICATION OF CURRENT TRANSFORMERS

A method of specifying current transformers for protective purposes is detailed in BS 3938. In this specification they are defined in terms of rated burden, accuracy class and accuracy limit.

Standard values of rated burden are:

2.5, 5, 7.5, 10, 15 and 30 VA.

Two accuracy classes are quoted 5P and 10P which gives a composite error at rated accuracy limit of 5% and 10% respectively.

Standard accuracy limit factors are:

5, 10, 15, 20 and 30.

The method of describing a current transformer is as follows:

15 VA Class 5P20

which means that it is rated for a burden 15 VA and will not have more than 5% error at 20 times rated current.

It is frequently more convenient to refer directly to the maximum useful voltage which can be obtained. In this connection, the knee-point of the magnetisation curve is defined as that point at which an increase of 10% of secondary voltage would increase the magnetising current by 50%. Design requirements for current transformers for general protective purposes are frequently specified in terms of knee-point voltage, magnetising current at the knee-point or at some other point, and secondary resistance. These are known in general as "Class X", current transformers.

PLATE 3.1 THE CT CHAMBER OF AN 11 kV CIRCUIT-BREAKER. The smaller set
of current transformers are for restricted earth-fault protection and
the larger for overcurrent protection. The earthed connection to the
bushing screen prevents a high voltage gradient between the
current transformer and earth (British Steel Corporation)

RATED SECONDARY CURRENT

Current transformers are usually designed to have rated secondary
currents of 0.5 A, 1 A or 5 A. Most burdens will require a definite
amount of VA at rated current and consequently will have an impedance which varies inversely as the square of the rated current, so
that the value of the rated secondary current does not appear to be
important. Many burdens, however, are situated at some distance
from the corresponding current transformers and, as the wire size of
the interconnecting leads is usually large enough to carry the current
produced by a current transformer of any secondary rating, the leads
introduce a definite resistance and therefore more burden at the
higher rated currents, e.g. lead resistance 1 ohm at 1 A correspond to
1 VA; lead resistance 1 Ω at 5 A corresponds to 25 VA. Clearly in all
cases where leads may be appreciable there is a great advantage in
using the lower rated current transformer. Modern practice favours
the use of the 1 A secondary windings.

SECONDARY WINDING IMPEDANCE

Bearing in mind the high value of secondary current which a protective current transformer may be required to deliver, it is desirable to make the secondary winding resistance as low as practicable to limit copper losses and therefore heating.

In the case of wound primary-type current transformers winding reactance also occurs, although its precise measurement and definition is a matter of some difficulty. Ring-type current transformers with a single symmetrical primary conductor and a uniformly distributed secondary winding have no secondary reactance.

PRIMARY WINDINGS

To achieve a reasonable output from a current transformer having a primary rating of 80 A or less would require a large core area and therefore it is more economical to increase the primary winding from a single turn to two, three or more turns. This of course necessitates an increase in secondary turns which increases knee-point voltage for a given core area. The additional primary turns may be attained by passing the primary conductor through a ring-type transformer a number of times or it may be a specially constructed transformer with a primary winding.

APPLICATION

In specifying current transformers the connected burden and mode of operation must be taken into account paying attention not only to the wide range of devices which may be connected, but also to the variation of impedance over the range of setting any relay. For example, the normal burden of an overcurrent relay is 3 VA at setting. The normal setting range of the relay is 50% to 200% of nominal current. Therefore a 1 A relay set to 50% would have a setting current of 0.5 A and the voltage across the coil at this current would be

$$V = \frac{3 \text{ VA}}{0.5 \text{ A}} = 6 \text{ V}$$

and the impedance would be

$$Z = \frac{6 \text{ V}}{0.5 \text{ A}} = 12 \text{ } \Omega.$$

At a setting of 200% the setting current would be 2 A, the voltage

$$V = \frac{3 \text{ VA}}{2 \text{ A}} = 1.5 \text{ V}$$

and the impedance

$$Z = \frac{1.5 \text{ V}}{2 \text{ A}} = 0.75 \ \Omega.$$

If the characteristic of the relay is to be maintained up to 20 times the relay setting, then a knee-point voltage not less than

$20 \times 6 \text{ V} \quad = 120 \text{ V}$ for a 50% setting
or $20 \times 1.5 \text{ V} = 30 \text{ V}$ for a 200% setting

would be required. The former is more onerous and therefore the lowest setting must be taken into account when specifying the knee-point voltage. There is, however, an alleviating factor in that a relay operating at 20 times its setting will have saturated magnetically and therefore the impedance will be reduced. The reduction for an overcurrent relay is about half the impedance at setting, which means that in the above case a knee-point voltage of 60 V would be satisfactory.

In many cases the current transformers associated with the over-current protection must also cater for earth-fault relays. An earth-fault relay having a minimum setting of 20% would have voltage at setting of

$$\frac{3 \text{ VA}}{0.2 \text{ A}} = 15 \text{ V and impedance of } \frac{15 \text{ V}}{0.2 \text{ A}} = 75 \ \Omega.$$

The maximum earth-fault level may be restricted to, say, twice the CT primary rating and therefore 10 times the relay setting. The knee-point voltage should therefore be greater than $10 \times 15 \text{ V} = 150 \text{ V}$, or allowing for saturation, 75 V.

In this case the size is determined by the earth-fault relay. A suitable current transformer would be a 7.5 VA Class 5P10. This would produce a voltage of 7.5 V at rated current when connected to a 7.5 Ω burden and would have only 5% error at 10 times rated current, i.e. at a voltage of $10 \times 7.5 \text{ V} = 75 \text{ V}$.

From the specification in the form 7.5 VA Class 5P10, the knee-point voltage can be estimated. If it has a 5 A secondary winding then at rated current it would produce 1.5 V across the rated burden and at 15 times rated current 22.5 V. As a rough guide the knee-point voltage is the product of the VA rating and the accuracy limit factor divided by the rated secondary current.

Class 5P is specified when phase-fault stability and accurate time grading is required. When these are unimportant, Class 10P is suitable.

It may be that more than one relay is to be connected to one set of current transformers in which case the total burden must be calculated. It is generally sufficient to add the burdens arithmetically but it should be borne in mind some alleviation may be available by adding the burden vectorially in case of difficulties in design.

It is not good engineering practice to specify a current transformer which is substantially larger than necessary as there is no advantage in performance and its cost would be higher and its dimensions greater.

EFFECT OF CT MAGNETISING CURRENT ON RELAY SETTING

The overall setting of a protection system is affected by the magnetising current of the current transformers and, whilst the effect may not be significant in the case of overcurrent relays, it can have some effect on the overall setting of an earth-fault relay and can sometimes have a profound effect on differential protection systems, particularly where a large number of current transformers are connected together. For example, a busbar zone protection scheme.

The primary operating current (P.O.C.) of a protection system is the sum of the relay setting current and the magnetising current of all the connected current transformers at the voltage across the relay at setting multiplied by the CT ratio.

QUADRATURE OR AIR-GAP CURRENT TRANSFORMERS

A quadrature or air-gap transformer is merely a current transformer with an air gap so that most of the primary ampere-turns are used to magnetise the core. This means that the flux, and therefore the secondary voltage, is proportional to primary current. More correctly, the secondary voltage is proportional to the rate of change of flux and therefore lags the primary current by 90°—hence the name quadrature current transformer.

SUMMATION CURRENT TRANSFORMER

There are two applications of the summation current transformer. One is the adding together the secondary current from a number of current transformers and is mainly used for measuring purposes. The other is used in pilot-wire protection systems to convert the inputs

PLATE 3.2   VOLTAGE TRANSFORMER IN THE SERVICE POSITION SHOWING
SECONDARY FUSE AND LINK. The cover has been removed from the
CT chamber beneath (British Steel Corporation)

from the current transformers in each phase to a single output for
comparison with a similar output from the remote end via the pilot
wires.

In the former case any input winding not in use must be left
open-circuited.

VOLTAGE TRANSFORMERS

The voltage transformer for use with protection has to fulfil only
one requirement, which is that the secondary voltage must be an
accurate representation of the primary voltage in both magnitude and
phase.

To meet this requirement, they are designed to operate at fairly low
flux densities so that the magnetising current, and therefore the ratio
and phase angle errors, is small. This means that the core area for a
given output is larger than that of a power transformer, which
increases the overall size of the unit. In addition, the normal three-
limbed construction of the power transformer is unsuitable as there

would be magnetic interference between phases. To avoid this inter-ference a five-limbed construction is used, which also increases the size. The nominal secondary voltage is sometimes 110 V but more usually 63.5 V per phase to produce a line voltage of 110 V.

ACCURACY

Only in a few of the many protection applications is the phase angle and ratio errors likely to be much significance. However, the likeli-hood of a voltage transformer being provided solely for protection is small and therefore the more stringent accuracies of instrumentation and metering are usually required.

All voltage transformers are required by BS 3941 to have ratio and phase-angle errors within prescribed limits over a 80% to 120% range of voltage and a range of burden from 25% to 100%.

For protection purposes, accuracy of measurement may be impor-tant during fault conditions when the voltage is greatly suppressed. Therefore a voltage transformer for protection must meet the extended range of requirements over a range of 5% to 80% rated voltage and, for certain applications, between 120% and 190% rated voltage.

PROTECTION

Voltage transformers are generally protected by HRC fuses on the primary side and fuses or a miniature circuit-breaker on the secondary side. As they are designed to operate at a low flux density their impedance is low and therefore a secondary side short-circuit will produce a fault current of many times rated current.

RESIDUAL CONNECTION

It is important that a voltage of the correct magnitude and phase angle is presented to directional earth-fault relays and the earth-fault elements of impedance relays. As an earth-fault can be any one of the three phases it is not possible to derive a voltage in the conventional manner. The solution is to use the residual or broken delta connection as shown in Fig. 3.5. Under three-phase balanced conditions the three voltages sum to zero. If one voltage is absent or reduced because of an earth-fault on that phase, then the difference between the normal voltage and that voltage is delivered to the relay. A secondary winding for this type of connection is in addition to the normal secondary winding.

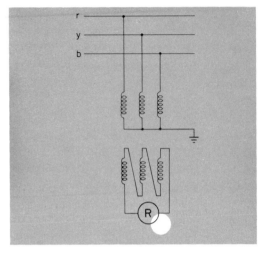

FIG. 3.5   BROKEN DELTA CONNECTION OF A VOLTAGE TRANSFORMER

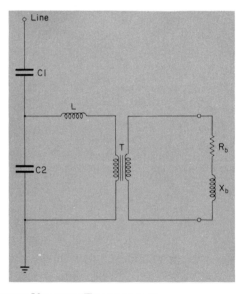

FIG. 3.6   CAPACITOR VOLTAGE TRANSFORMER

CAPACITOR VOLTAGE TRANSFORMERS

At voltages of 132 kV or more, the cost of electromagnetic voltage transformers is very high. A more economical proposition is the capacitor voltage transformer. This is virtually a capacitance voltage divider with a tuning inductance and an auxiliary transformer as shown in Fig. 3.6.

Any simple voltage-divider system suffers from the disadvantages that the output voltage varies considerably with burden. If, however, C2 is tuned with a reactor, the burden can be varied over a wide range with very low regulation. It is not feasible to produce directly the usual 63.5 V as C2 would be impossibly large and therefore a potential of around 12 kV is developed across C2. This is applied to an electromagnetic unit and the 63.5 V derived from its secondary winding. This method also has the advantage that a tapped winding can be provided to accommodate the fairly wide tolerances of capacitors.

Chapter 4

Fault Calculations

In order to predict the performance of a protection scheme it is necessary to know what the fault conditions will be. Although some relays will be required to deal with overloads, undervoltages, etc., the majority will be concerned with the detection of short-circuit conditions. To determine the fault level when a short-circuit occurs requires a knowledge of the impedance of the various components of the power system and the ability to calculate the current in every part of the system.

IMPEDANCE

Although an impedance consists of a resistance and a reactance it is usually sufficient to take only the reactance into consideration in fault calculations. If a computer is used for the calculation it is just as simple to include resistance but if other means are used its inclusion is an unnecessary complication. In most cases the exclusion of resistance is justified in that the resistance is only a small fraction of the impedance and even if it were as high as 20% it would only change the impedance by about 2%.

The exception is in cables where, if the cross-sectional area is small, the resistance is of the same order as the reactance. However, as cables have a very low impedance compared to transformers and generators the overall effect of ignoring resistance is small.

FAULT LEVEL

When evaluating relay performance it is usual to use the three-phase fault level and, if earth-fault relays are involved, the earth-fault level. It is appreciated that a phase-phase fault is far more likely than a three-phase fault; however, the three-phase value is used on the basis that it is the most onerous condition.

Calculation of a three-phase fault is fairly straightforward as it is a balanced fault. That is, the current in each of the three phases has the same magnitude and they are 120° apart. Therefore all that is required is to calculate the current in one phase using the phase-neutral voltage and the impedance per phase. For example, an 11 kV generator has an impedance of 1.61 Ω/phase:

$$\text{phase voltage } \frac{11,000}{\sqrt{3}} = 6350 \text{ V},$$

$$\text{fault current} = \frac{6350}{1.61} = 3970 \text{ A}.$$

Although current is used in determining relay settings it is more usual to perform fault calculations in MVA as this avoids complications when there is a change in voltage, i.e. when transformers are involved. Therefore, fault level

$$= 3 \times 6350 \times 3970 \times 10^{-6} = 75 \text{ MVA},$$
$$\text{or } \sqrt{3} \times 11 \text{ kV} \times 3.97 \text{ kA} = 75 \text{ MVA}.$$

A quicker way would be to perform the calculation in one operation,

$$\text{viz. } 3 \times 11,000 \times \frac{11,000}{3 \times 1.61} \times 10^{-6} = 75 \text{ MVA}$$

$$\text{or in symbols } \frac{3 \, V \times V \times 10^{-6}}{3Z} = \frac{V^2}{Z} \times 10^{-6}$$

or if V is in kV

$$\text{fault MVA} = \frac{V^2}{Z}.$$

If the generator was rated as 15 MW, 0.8 power factor then the rating would be:

$$\frac{15}{0.8} = 18.75 \text{ MVA}.$$

The rating as a fraction of fault level would be

$$\frac{18.75}{75} = \tfrac{1}{4} \text{ or } 25\%.$$

This ratio is known as the percentage impedance or $Z\%$. Generator and transformer impedances are generally expressed in this way

$$Z\% = \frac{\text{MVA rating}}{\text{fault level}} \times 100\% = \frac{\text{MVA rating}}{V^2} \times Z \times 100\%.$$

Check

$$Z\% = \frac{18.75}{11^2} \times 1.61 \times 100\% = 25\%.$$

$$\text{Fault level} = \frac{18.75}{25\%} \times 100\% = 75 \text{ MVA}.$$

Another example—Generator 12.5 MVA, 20%:

$$\text{three-phase fault level} = \frac{12.5}{20\%} \times 100\% = 62.5 \text{ MVA}.$$

Rather than calculate in percentages, and multiply by 100% every time, it is convenient to use per unit values. For instance, 20% means every 20 in 100 and it could be written 0.2 p.u., i.e. 0.2 in every 1. Therefore, for the above generator Zp.u. = 0.2:

$$\text{three-phase fault level} = \frac{12.5}{0.2} = 62.5 \text{ MVA}$$

or the previous generator Zp.u. = 0.25:

$$\text{three-phase fault level} = \frac{18.75}{0.25} = 75 \text{ MVA}.$$

It is also more convenient to convert all per unit impedances to a common base, say 10 MVA, in the following manner:

$$18.75\text{-MVA generator } Z\text{p.u.} = \frac{10}{18.75} \times 0.25 = 0.133 \text{ p.u.},$$

$$12.5\text{-MVA generator } Z\text{p.u.} = \frac{10}{12.5} \times 0.2 = 0.16 \text{ p.u.}$$

The reason for this is so that the relative values of impedance can be attributed to every component in the circuit and therefore allow easy calculations.

If a transformer rated at 4 MVA having an impedance of 6% is connected to the 18.75 MVA generator and both impedances are converted to a base of 10 MVA, then the generator impedance is 0.133 p.u. and the transformer is

$$\frac{10}{4} \times \frac{6\%}{100\%} = 0.15 \text{ p.u.}$$

The fault level on the secondary side of the transformer is

$$\frac{10 \text{ MVA}}{0.133 + 0.15} = \frac{10}{0.283} = 35.3 \text{ MVA}.$$

If there were two 4 MVA transformers in parallel each having an impedance of 6% then the total current impedance for a fault on the secondary side would be

$$0.133 + \frac{0.15}{2} = 0.208 \,\text{p.u.}$$

and the fault level would be

$$\frac{10 \,\text{MVA}}{0.208} = 48 \,\text{MVA}.$$

If one of the transformers instead of being 4 MVA was 3 MVA with an impedance of 6% then the system would be as Fig. 4.1:

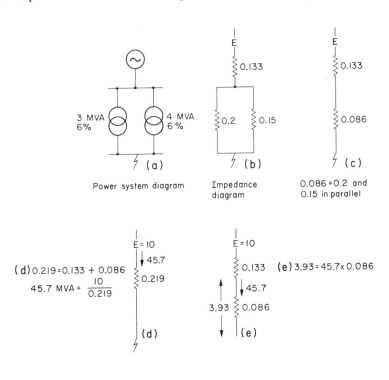

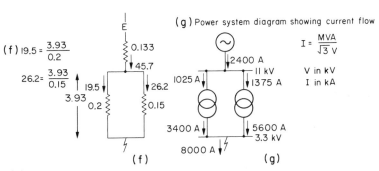

FIG. 4.1

Zgen = 0.133 p.u.

Z_{T1} = 0.15 p.u.

$$Z_{T2} = \frac{10}{3} \times \frac{6}{100} = 0.2 \text{ p.u.}$$

The two transformers in parallel

$$\frac{1}{Z} = \frac{1}{0.15} + \frac{1}{0.2},$$

Z = 0.086 p.u.

and the fault level

$$\frac{10}{0.133 + 0.086} = 45.7 \text{ MVA.}$$

By the application of Ohm's law the fault current for any power system can be calculated by constructing an impedance network in which all the components are represented by a per unit impedance and the fault level is the "current" which is determined by dividing the MVA base—the "voltage"—by the per unit impedance.

In the example the fault level is 45.7 MVA. Across the two transformers in parallel the "voltage" is 45.7 × 0.086 = 3.93 and therefore the contribution to the fault through the 4 MVA transformer is

$$\frac{3.93}{0.15} = 26.2 \text{ MVA}$$

and through the 3 MVA transformer

$$\frac{3.93}{0.2} = 19.5 \text{ MVA.}$$

Figure 4.1 shows the steps of calculation starting with the system diagram with reactances at (a), the impedance diagram at (b), the circuit reduction at (c) and (d) and the establishment of MVA flow at (e) and (f) culminating in the system diagram with current flow at (g) calculated from $I = \text{MVA}/\sqrt{3}V$. In an actual calculation some of these steps would be omitted but the object remains the same. No matter how complicated the network is, the object is to reduce it to a single impedance from which the fault MVA and its flow in various parts of the circuit is determined so that the performance of the protection can be predicted.

The elements of a power system are specified as follows:

(a) Generators and Transformers—per cent impedance on rating.
(b) Feeders and Interconnectors—actual impedance/phase.
(c) Reactors—voltage drop at rated current.

To convert these to per unit values on a common base

(a) $Zp.u. = \dfrac{Z\%}{100} \times \dfrac{MVA\ base}{MVA\ rating}$,

(b) $Zp.u. = Z \times \dfrac{MVA\ base}{V^2}$,

(c) $Zp.u. = \dfrac{V_R}{I_R} \times \dfrac{MVA\ base}{V^2}$.

Figure 4.2 shows a diagram with all these power system components.
Using a 10 MVA base:

Generator   20 MVA, 25%

$$X_g = \frac{25}{100} \times \frac{10}{20} = 0.125 \text{ p.u.}$$

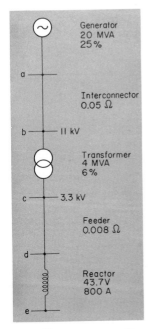

FIG. 4.2   TYPICAL IMPEDANCE VALUES OF THE COMPONENTS OF A POWER
SYSTEM

Interconnector   0.05 Ω, 11 kV

$$X_I \times \frac{10}{11^2} \times 0.05 = 0.004 \, \text{p.u.}$$

Transformer   4 MVA, 6%

$$X_T = \frac{6}{100} \times \frac{10}{4} = 0.15 \, \text{p.u.}$$

Feeder   0.008 Ω, 3.3 kV

$$X_F = \frac{10}{3.3^2} \times 0.008 = 0.007 \, \text{p.u.}$$

Reactor   43.7 V, 800 A

$$X_R = \frac{10}{3.3^2} \times \frac{43.7}{800} = 0.05 \, \text{p.u.}$$

Fault at a $= \dfrac{10}{0.125} = 80 \, \text{MVA}$,

$$b = \frac{10}{0.125 + 0.004} = \frac{10}{0.129} = 77.5 \, \text{MVA},$$

$$c = \frac{10}{0.129 + 0.15} = \frac{10}{0.279} = 35.8 \, \text{MVA},$$

$$d = \frac{10}{0.279 + 0.007} = \frac{10}{0.286} = 35 \, \text{MVA},$$

$$e = \frac{10}{0.286 + 0.05} = \frac{10}{0.336} = 29.8 \, \text{MVA}.$$

Typical impedance values can be attributed to all components of a power system in the absence of definite information. Transformer impedances are usually easy to determine as the value is marked on the rating plate. The impedance of generators is usually of secondary importance as most distribution systems generally have a much higher infeed and fault contribution from the public electricity-supply system. There is, however, a continuing increase in offshore installations which there is no external supply. In this case the performance of the generator is of prime importance.

GENERATORS

The performance of a generator under fault conditions is more complicated than that of any other part of the distribution system. The fault current is initially about 8 times full-load current decaying rapidly to 5 times full-load current and then decaying less rapidly to less than full-load current. The three stages are known as sub-transient, transient and synchronous respectively.

The synchronous or steady-state reactance of a generator is high because of armature reaction and is in the range of 1.5 to 2.5 p.u. at the machine rating. The value used is made up of two components the actual reactance of the machine which is small and a fictitious reactance. When a fault occurs the current lags the voltage by 90° and the position of the field is such that it is demagnetised by the current flowing in the stator conductors so that the air-gap flux and therefore the generated e.m.f. is low. Rather than use this low e.m.f. and calculate the fault current by dividing it by the actual reactance it is more convenient to use the initial e.m.f., E, and divide it by a fictitious value to obtain the same result.

The rapid change of flux due to the demagnetising effect of the stator current results in an induced current in the field which opposes the change and tends to maintain the field flux. Thus the initial flux, e.m.f. and fault current are somewhat higher than the steady-state value, and decay exponentially towards the steady-state value. Once again a fictitious value of reactance coupled with the e.m.f. is used in calculation—the reactance being termed the transient value.

There is one other effect and that is the damper winding in the pole face will also produce a flux opposing demagnetisation and will result in a fault current slightly higher than that produced under transient conditions. This fault current is of very short duration, it decays exponentially, and the fictitious value associated with it is known as the subtransient reactance.

The reactance values associated with a generator are typically:

Subtransient reactance X_d'' value 0.12 p.u.
Transient reactance X_d' value 0.16 p.u.
Synchronous reactance X_d value 2 p.u.

The e.m.f. at no load would be the same as the system voltage, V, which at the nominal value is 1. At any other load the e.m.f. would be greater:

$$E'' = [(V + X_d'' I \sin \phi)^2 + (X_d'' I \cos \phi)^2]^{1/2},$$
$$E' = [(V + X_d' I \sin \phi)^2 + (X_d' I \cos \phi)^2]^{1/2},$$
$$E = [(V + X_d I \sin \phi)^2 + (X_d I \cos \phi)^2]^{1/2},$$

at normal system voltage where I is the p.u. value of load = 1 at rated MVA and $\cos \phi$ is the power factor of the load.

$E'' = [(1 \times X_d''I \sin \phi)^2 + (X_d''I \cos \phi)^2]^{1/2}$,

for the value given at full load 0.8 p.f.

$E'' = [(1 + 0.12 \times 1 \times 0.6)^2 = (0.12 \times 1 \times 0.8)^2]^{1/2}$,
$E'' = 1.076$.

Subtransient current (or MVA) $= \dfrac{1.076}{0.12} = 8.97 \times$ FL current

or $8.97 \times$ rated MVA.

Another example:

at 70% load, 0.9 p.f.,
$\cos \phi = 0.9$, $\phi = 25.8°$,
$\sin \phi = 0.436$,
$E' = [(1 + 0.16 \times 0.7 \times 0.436)^2 + (0.16 \times 0.7 \times 0.9)^2]^{1/2}$,
$E' = 1.054$.

Transient current (or MVA) $= \dfrac{1.054}{0.16} = 6.58 \times$ FL current

or $6.58 \times$ rated MVA.

The above are the initial values of the current under short-circuit conditions. The subtransient value would disappear in a fraction of a second whilst it would take several seconds for the transient value to decay.

The decay is exponential and typical time constants are

$T_{d0}'' = 0.1$ s,
$T_{d0}' = 5$ s.

These are open-circuit time constants, under short-circuit conditions the value is modified as follows:

$$T_d'' = \frac{X_d''}{X_d} T_{d0}'',$$

$$T_d' = \frac{X_d'}{X_d} T_{d0}'.$$

These would be the time constants for a terminal fault. If the fault was on the secondary side of a transformer, reactance X_T, connected to the generator then the short-circuit time constants would be

$$T_d'' = \frac{X_d'' + X_T}{X_d + X_t} T_{d0}'',$$

$$T_d' = \frac{X_d' + X_T}{X_d + X_T} T_{d0}'.$$

Example
15 MW generator power factor 0.8:

$$\frac{15}{0.8} = 18.75 \text{ MVA},$$

$X_d'' = 0.12$ p.u.,
$X_d' = 0.25$ p.u.,
$X_d = 2.0$ p.u.,
$T_d'' = 0.15$ s,
$T_d' = 5.0$ s,

operating at full load prior to fault

Power factor $= \cos \phi = 0.8$,
$\phi = 36.9°$,
$\sin \phi = 0.6$,

As the only item involved is the generator, a base of 18.75 MVA, the generator rating, can be used

$E'' = [(1 + 0.12 \times 1 \times 0.6)^2 = (0.12 \times 1 \times 0.8)^2]^{1/2}$
$\qquad = 1.076,$
$E' = [(1 + 0.25 \times 1 \times 0.6)^2 + (0.25 \times 1 \times 0.8)^2]^{1/2}$
$\qquad = 1.167,$
$E = [(1 + 2 \times 1 \times 0.6)^2 + (2 \times 1 \times 0.8)^2]^{1/2}$
$\qquad = 2.72,$

$$I'' = \frac{1.076}{0.12} = 8.97 \times \text{FL current},$$

$$I' = \frac{1.167}{0.25} = 4.67 \times \text{FL current},$$

$$I = \frac{2.72}{2} = 1.36 \times \text{FL current},$$

$$T_d'' = \frac{0.12}{2} \times 0.1 = 0.006 \text{ s},$$

$$T_d' = \frac{0.25}{2} \times 5 = 0.625 \text{ s},$$

$$\text{FL current} = \frac{18.75}{\sqrt{3} \times 11} = 984 \text{ A}.$$

When $t = 0$

$I = 8.97 \times 984 = 9684$ A.

This decreases to the transient value in about 0.03 s, i.e. 4.67×984 = 4595 A. The final value is $1.36 \times 984 = 1338$ A. Therefore the part which is decaying is $4595 - 1338 = 3257$ A.

When $t = 0.1$ s

$$i = 3257 \, e^{-(0.1/0.625)} + 1338 = 4113 \text{ A}.$$

When $t = 0.2$ s

$$i = 3257 \, e^{-(0.2/0.625)} = 1338 = 3703 \text{ A}.$$

When $t =$	0.4	0.6	0.8	1.0	2.0	∞
$i =$	3055	2585	2244	1999	1471	1338

The subtransient current is of interest only to the switchgear designer to determine closing duty. As far as protection is concerned it has disappeared before any relay operation.

Both the transient and synchronous values are used to determine the performance of the protection. The transient value for high-speed and instantaneous schemes and the synchronous value for any scheme that has a time delay.

If the generator was operating at no-load prior to the fault then the three values would be

Subtransient $\quad I'' = \dfrac{1}{0.12} \times 984 = 8200 \text{ A}$

Transient $\quad I' = \dfrac{1}{0.25} \times 984 = 3936 \text{ A}$

Synchronous $\quad I = \dfrac{1}{2} \times 984 = 482 \text{ A}$

Although there is some reduction throughout the biggest effect is, of course, under synchronous conditions.

In practice as soon as the fault occurs there would be a reduction in voltage which would cause the automatic voltage regulator to increase the field thus increasing the synchronous value from no-load to the value for full load—Fig. 4.3 shows the values calculated above in a graphical form.

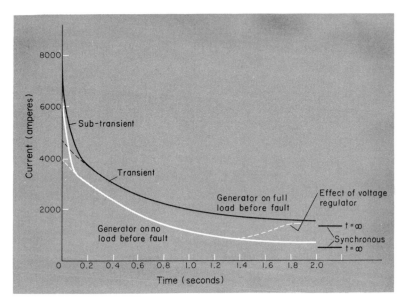

FIG. 4.3   THE PERFORMANCE OF A 15-MW GENERATOR UNDER FAULT
CONDITIONS

CABLES

The resistance of a cable is determined by the cross-sectional area
of the conductors but the reactance depends on the distance between
the conductors, i.e. the insulation thickness which depends on the
voltage. The inductance can be calculated from

$$L = 0.46 \log \frac{d}{r} \ \mu\text{H/m},$$

where d is the distance between conductor centres and r is the
conductor geometric mean radius. It should be remembered that the
cable made from a number of strands and the radius which is
calculated from area $= \pi r^2$ is not the geometric mean radius which is
approximately 78% of that value. Also the actual radius which is used
in conjunction with the insulation thickness to determine d is 15%
larger than the calculated value.

Example: 400 mm² cable three-core screened, 6350/11,000 V.

$$r_e = \left(\frac{400}{\pi}\right)^{1/2} = 11.28 \text{ mm},$$

$$r = 0.78 \times 11.28 = 8.8,$$

insulation thickness 5.6 mm

$d = (1.15 \times 11.28)2 + 5.6 = 33.8 \text{ mm},$

$L = 0.46 \log \dfrac{33.8}{8.8} = 0.255 \ \mu\text{H/m},$

and the reactance at 50 Hz

$X = 2\pi f L = 2\pi \times 50 \times 0.255$
$= 80 \ \mu\Omega/\text{m}.$

Where three single-core cables are used there is an increase in reactance because the distance between the conductors is increased. A 400 mm^2 single-core cable has an overall diameter of 39 mm and therefore if the three cables are mounted in trefoil formation (Fig. 4.3(a)), then

$L = 0.46 \log \dfrac{39}{8.8} = 0.297 \ \mu\text{H/m}$

and the reactance

$X = 93.4 \ \mu\Omega/\text{m}.$

If the cables are laid flat as in Fig. 4.4(b) then d is the geometric mean distance which is

$d = (d_1 d_2 d_3)^{1/3} = (39 \times 39 \times 39 \times 2)^{1/3}$
$= 49.2 \text{ mm},$

$L = 0.46 \log \dfrac{49.2}{8.8} = 0.343 \ \mu\text{H/m},$

$X = 108 \ \mu\Omega/\text{m}.$

There is no need to calculate the value in every instance, a close approximation can be obtained by using typical values.

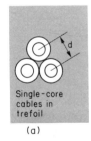

Single-core
cables in
trefoil

(a)

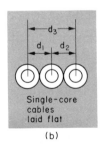

Single-core
cables
laid flat

(b)

FIG. 4.4

Reactance in $\mu\Omega$/m

	Three-core	Trefoil	Flat
11 kV	80	95	110
415 V	75	87	100

From the actual reactance the per unit reactance at the chosen base can be calculated from

$$Z\text{p.u.} = \frac{\text{MVA base}}{V^2} \times Z \quad \text{(where } V \text{ is in kV)}$$

for example:
1 km, 11 kV, three-core cable

$$X = 1000 \times 80 \times 10^{-6} = 0.08\ \Omega,$$

$$X\text{p.u.} = \frac{10}{11^2} \times 0.08 = 0.0066\ \text{p.u.}$$

15 m, 415 V, three-core cable

$$X = 15 \times 75 \times 10^{-6} = 0.001\ 125\ \Omega,$$

$$X\text{p.u.} = \frac{10}{.415^2} \times 0.001\ 125 = 0.065\ \text{p.u.}$$

Comparison of the per unit reactance values shows that the 415 V cable will have a much greater effect on the fault current than the much longer 11 kV cable.

SOURCE IMPEDANCE

This is merely a value which represents the impedance between the system under consideration and the source. The value is determined by the fault level at the incoming busbar. If the actual fault level is not known then a value based on the switchgear rupturing capacity is used. For example, if the fault level or rupturing capacity is 250 MVA, then the source impedance on a 10 MVA base is

$$\frac{10}{250} = 0.04\ \text{p.u.}$$

Figure 4.5(a) shows part of a typical distribution system and Fig. 4.5(b) the impedance diagram. As can be seen an impedance of 0.04 has been included to limit the fault level at the plant substation 11 kV busbars to 250 MVA.

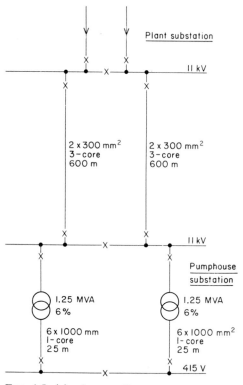

FIG. 4.5. (a)   SYSTEM DIAGRAM

Plant substation

11 kV

2 x 300 mm²
3 – core
600 m

2 x 300 mm²
3 – core
600 m

11 kV

Pumphouse
substation

1.25 MVA
6 %

1.25 MVA
6 %

6 x 1000 mm
1 – core
25 m

6 x 1000 mm²
1 – core
25 m

415 V

E = 10

0.04

0.002

0.002

0.48

0.48

0.0726

0.0726

FIG. 4.5. (b)
IMPEDANCE DIAGRAM

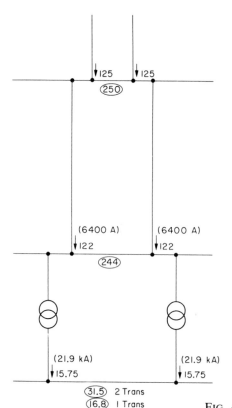

250

↓125   ↓125

(6400 A)   (6400 A)
↓122   ↓122

244

(21.9 kA)   (21.9 kA)
↓15.75   ↓15.75

31.5   2 Trans
16.8   1 Trans

FIG. 4.5. (c)   MVA AND CURRENT FLOW

MOTORS

There is also a contribution to a fault from any induction motors which are connected at the time of the fault. The initial value will be roughly equal to the motor-starting current but will decay rapidly to zero. It is mainly of interest to switchgear and power system designers as the effect on differential protection is small and the current will have disappeared by the time overcurrent relays operate. It could affect the operation of fast-acting devices such as fuses or miniature circuit-breakers but the accuracy of these devices is not of a high order and therefore a precise appraisal is unnecessary.

Synchronous motors behave in the same way as generators, the fault passing through the subtransient, transient to the synchronous stage.

PRACTICAL EXAMPLE

Figure 4.5(a) shows part of a typical distribution system. There is an incoming 11 kV supply to the plant substation. Two 11 kV interconnectors to the pump house substation where there are two 11/0.415 kV transformers.

The first step is to construct an impedance diagram to a common base, say 10 MVA. The fault level at the plant substation 11 kV busbars is 250 MVA and therefore the source impedance

$$X_s = \frac{10}{250} = 0.04 \text{ p.u.}$$

The two interconnectors are each two 300 mm^2 cables in parallel and therefore the reactance of each interconnector is

$$\tfrac{1}{2} \times 600 \times 80 \times 10^{-6} = 0.024 \ \Omega$$

$$X_I = 0.024 \times \frac{10}{11^2} = 0.002 \text{ p.u.}$$

The two 1.25 MVA transformers are each

$$X_T = \frac{10}{1.25} \times \frac{6}{100} = 0.48 \text{ p.u.}$$

The interconnecting cables to the 415 V switchgear are assumed to be flat in configuration:

$$\tfrac{1}{2} \times 25 \times 100 \times 10^{-6} = 0.001\ 25 \ \Omega$$

$$X_C = 0.001\ 25 \times \frac{10}{0.415^2} = 0.0726 \text{ p.u.}$$

From this diagram the fault level at any particular part can be determined. In more complicated arrangements it may be necessary to calculate the combined impedance of various parts of the system and redraw the impedance diagram to simplify it to the extent where the calculation is straightforward. It may be that more than one redraw is necessary before the calculation can be made.

Returning to the impedance diagram of the system shown in Fig. 4.5(b):

a fault at the plant substation 11 kV busbar is

$$\frac{10}{0.04} = 250 \, \text{MVA}, \text{ of course,}$$

a fault at the pump house substation 11 kV busbar is

$$\frac{10}{0.04 + (0.002)\frac{1}{2}} = \frac{10}{0.041} = 244 \, \text{MVA},$$

a fault at the 415 V busbar is

$$\frac{10}{0.041 + (0.48 + 0.0726)\frac{1}{2}} = \frac{10}{0.3173} = 31.5 \, \text{MVA}$$

with two transformers, and

$$\frac{10}{0.041 + 0.48 + 0.0726} = \frac{10}{0.5936} = 16.8 \, \text{MVA}$$

with one transformer.

Note that if only one transformer is connected the current per transformer is greater than if two transformers were connected. The system diagram and the flow through the various parts of the system is as shown in Fig. 4.5(c). Alternatively the actual current flow, marked in brackets, can be shown.

In the case of Fig. 4.6 the situation is a little more complicated. The impedance diagram is drawn and the reactance calculated as in the previous example.

The complication is the delta-connection of the 0.002 and the two 0.0015 impedances. However, as these are low compared to the source and generator impedances the generation can be assumed to be connected to the fault busbar. This means that the fault level can be calculated at any busbar without taking the generator into account and adding the contribution from the generator, in this case 10/0.133 = 75 MVA, to the results.

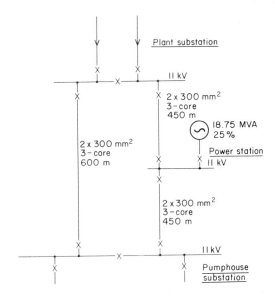

FIG. 4.6. (a)   SYSTEM DIAGRAM

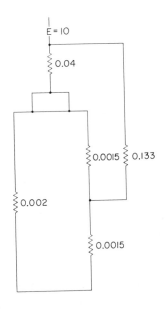

FIG. 4.6. (b)   IMPEDANCE DIAGRAM

For example, a fault at the plant substation would be $10/0.04 = 250$ MVA plus the generator contribution of 75 MVA, i.e.

$250 + 75 = 325$ MVA, and at the pumphouse substation

0.002 in parallel with $0.0015 + 0.0015$

$$\frac{1}{X} = \frac{1}{0.002} + \frac{1}{0.003} = \frac{1}{0.0012},$$

add the source impedance

$0.0012 + 0.04 = 0.0412$ p.u.

$$\text{Fault level} = \frac{10}{0.0412} + 75 = 318 \text{ MVA}.$$

If, however, the delta-connected circuit is not small as in the case shown in Fig. 4.7 quite an elaborated calculation using a delta star conversion is required.

The part of the power system shown is a power station with two incoming supplies and an interbusbar reactor which is switched in when the generators are in operation to keep the fault level to 250 MVA which is the rupturing capacity of the switchgear.

The per unit reactances are calculated as usual and the impedance diagram drawn. As can be seen there is no simple series or parallel combination which can be eliminated and so a delta star conversion must be made to the 0.07, 0.07 and 0.045 impedances. The equation for the conversion is

$$Z_a = \frac{Z_1 Z_2}{Z_1 + Z_2 + Z_3}.$$

Note that if the star was superimposed on the delta Z_a would lie between Z_1 and Z_2.

$$\text{Hence } Z_b = \frac{Z_3 Z_1}{Z_1 + Z_2 + Z_3}$$

$$\text{and } \quad Z_c = \frac{Z_3 Z_1}{Z_1 + Z_2 + Z_3},$$

in this particular case

$$Z_a = \frac{0.07 \times 0.07}{0.07 + 0.07 + 0.045} = 0.0265,$$

$$Z_b = Z_c = \frac{0.045 \times 0.07}{0.185} = 0.017.$$

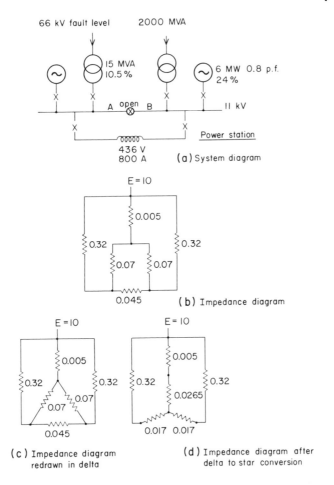

FIG. 4.7

Figure 4.7(d) shows that there are now series and parallel combinations which can be simplified and the fault level calculated. Figure 4.8(a) shows the impedance diagram redrawn for a fault on busbar B. This series/parallel circuit could be easily solved by most calculators as follows:

0.32 $\boxed{+}$ 0.017 $\boxed{=}$ $\boxed{1/x}$ $\boxed{+M}$ 0.005 $\boxed{+}$ 0.0265 $\boxed{=}$ $\boxed{1/x}$ $\boxed{+M}$

RM $\boxed{1/x}$ $\boxed{+}$ 0.017 $\boxed{=}$ $\boxed{1/x}$ $\boxed{MC}$ $\boxed{+M}$ 0.32 $\boxed{1/x}$ $\boxed{+M}$ $\boxed{RM}$

The final $\boxed{RM}$ multiplied by the base MVA will give the fault level, or using the longhand method

0.32 + 0.017 = 0.337,   Fig. 4.8(a)
0.005 + 0.0265 = 0.0315.

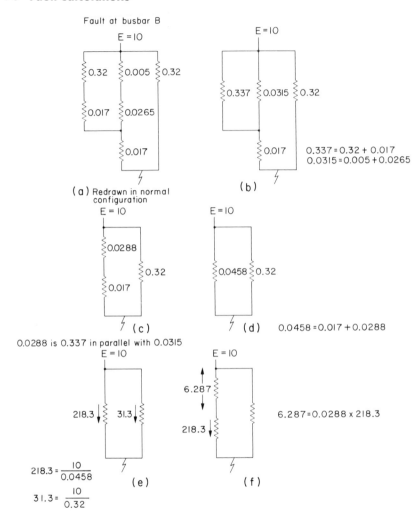

FIG. 4.8

These two in parallel

$$\frac{1}{0.337} + \frac{1}{0.0315} = 34.71,$$ Fig. 4.8(b)

$$\frac{1}{34.71} = 0.0288,$$

in series with 0.017

$$0.0288 + 0.017 = 0.0458.$$ Fig. 4.8(c)

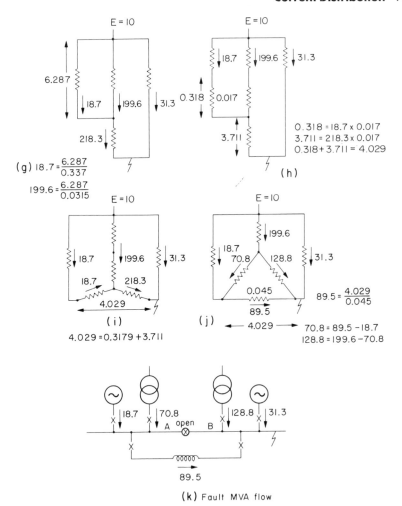

(k) Fault MVA flow

Fault current $= \dfrac{10}{0.0458} = 218.3$ MVA   Fig. 4.8(d)

plus
Fault current from generator

$\dfrac{10}{0.32} = 31.3$ MVA   Fig. 4.8(d)

Total fault current $= 218.3 + 31.3 = 249.6$ MVA.

CURRENT DISTRIBUTION

Figure 4.8(e) shows the current distribution. The "voltage" across circuits A and B in parallel is $218.3 \times 0.0288 = 6.287$.

"Current" in A $= \dfrac{6.287}{0.337} = 18.7$ MVA,
Fig. 4.8(f)
Fig. 4.8(g)

"Current" in B $= \dfrac{6.287}{0.0315} = 199.6$ MVA.

"Current" distribution in the equivalent star circuit is shown in Fig. 4.8(i). The voltage across the bottom of the star is

$18.7 \times 0.017 + 218.3 \times 0.017 = 4.029.$ Fig. 4.8(i)

The "current" in the reactor is this "voltage" divided by the impedance of the reactor.

"Current" in reactor $= \dfrac{4.029}{0.045} = 89.5$ MVA   Fig. 4.8(j)

which determines the other values of "current" in the delta and therefore the MVA distribution is as shown in Fig. 4.8(k).

The method has been made rather more elaborate than necessary in order to show every step in the calculation. Many of the intermediate steps would be omitted.

From Fig. 4.8(k) the actual fault current should be calculated as it is to this that the protection responds.

EARTH FAULTS

The earth-fault level of a distribution system is determined by the method by which it is earthed. Although earthing at each substation is by means of electrodes driven into the ground, very little of the earth-fault current flows via this route.

In distribution system at the higher voltages, i.e. 33 kV, 11 kV and 6.6 kV, the main earth-fault current flow is via the cable sheath and armouring whereas at the utilisation voltages of 3.3 kV and below the main earth fault is usually a direct bonded conductor from the equipment to the distribution transformer.

Because the cable sheath and armouring are used on the higher voltages the earth-fault path has a higher impedance than if it was directly bonded. This means that for an earth fault at a location removed from the substation where the distribution transformer is installed a higher proportion of the voltage will be dropped in the return path of a value such that the voltage at the fault would be unacceptably high. It is for this reason that all metalwork at each location must be earthed. When this is done it means that the whole area is at high voltage and as such does not constitute a danger. It does, however, stress the insulation of any connections between the fault area and the distribution point, e.g. pilot wires or telecommunications circuits. Because of the latter there is a requirement that the rise of earth voltage shall not exceed 430 V at any point. To meet this requirement, in general, requires that 33 kV, 11 kV and 6.6 kV systems be earthed via a neutral earthing resistor.

The effect of a neutral earthing resistor is to limit the earth-fault current to a relatively low value, which means that during an earth fault most of the phase/neutral voltage is dropped across this resistor. The reduction in fault current is also necessary so that the earth-fault current does not exceed the current-carrying capability of the sheath and armour. Table 4.1 gives typical resistance values for cable sheath and armour.

TABLE 4.1. RESISTANCE OF LEAD SHEATH AND STEEL WIRE ARMOUR FOR THREE-CORE PILCSWA, 6350/11,000 V CABLE

Conductor size (mm^2)	Resistance ($\mu\Omega$/m)		
	Lead sheath	Steel armour	Combined
50	1250	700	450
70	1050	650	400
95	950	600	370
120	870	570	340
150	760	540	310
185	700	500	290
240	570	460	250
300	500	340	200
400	430	310	180

The value of resistance chosen for the neutral earthing resistor is such that the earth-fault current is limited to around the full-load current of the transformer and so, for the purpose of assessing relay performance, it can be assumed that this will be the earth-fault level of the whole system.

Usually each transformer will have its own neutral earthing resistor and all transformers in a group must be earthed. There will be, therefore, an earth-fault level throughout the system which, as far as the protection is concerned, is dependent only on the number of transformers connected.

The earth-fault level of a 415 V system is almost indeterminate The rupturing capacity of the switchgear is usually 31 MVA which is a maximum fault current of 43,000 A and yet if the fault path has an impedance of only 0.1 Ω the fault current is reduced to almost a twentieth of that value.

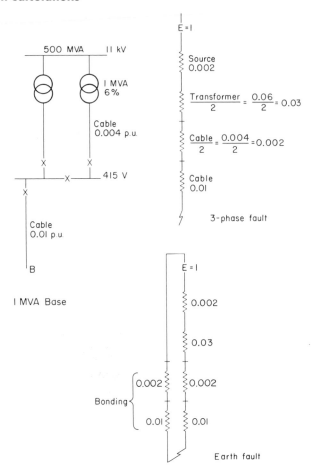

FIG. 4.9   IMPEDANCE DIAGRAM SHOWING EARTH RETURN IMPEDANCE

As shown earlier cable impedance does have a large effect on the fault level at 415 V and a sufficiently accurate value of earth-fault level can be obtained if the reactance of the return path is assumed to be the same as the cable reactance from the transformer to the fault. In other words, in the system shown in Fig. 4.9 for a fault at B the three-phase fault level would be:

$$\frac{1}{0.002 + \frac{1}{2}(0.06 + 0.004) + 0.01} = \frac{1}{0.044} = 22.7 \text{ MVA } (31.6 \text{ kA})$$

and the earth-fault level approximately

$$\frac{1}{0.044 + \frac{1}{2}(0.004) + 0.01} = \frac{1}{0.056} = 17.9 \text{ MVA } (24.9 \text{ kA}).$$

Chapter 5

Time-graded Overcurrent Protection

The induction relay is used in the great majority of time-graded overcurrent protection schemes. Its construction may vary but its characteristic and method of setting are common to all types.

The basic requirement is that firstly, it has to have an inverse time/current characteristic; that is it has a long operating time at low multiples of setting current and a shorter operating time at high multiples of setting current. Secondly, it must have the means of adjusting the current setting and the time of operation at a given multiple of setting.

The standard relay has a characteristic

$$t = 3(\log M)^{-1} \text{ or } \frac{3}{\log M}$$

where M is the multiple of setting. This means that at twice setting current, operating time $t = 10$ and at 10 times the setting current, operating time $t = 3$. Figure 5.1 shows the characteristic time/current curve.

In order to adjust the current setting the relay coil is arranged to have a tapped winding which is connected to a plugbridge. With the plug in the first position the whole of the coil is in circuit and therefore the relay is most sensitive. In the seventh position only one-quarter of the coil is in circuit and therefore it requires 4 times as much current to produce the same response. The first tap is generally arranged to give a relay setting of 50% of relay rating and therefore the seventh tap would be $4 \times 50 = 200\%$. Figure 5.2 shows typical connections.

The seven plugbridge positions would be marked 50%, 75%, 100%, 125%, 150%, 175% and 200%. Figure 5.3 shows the response of the relay to various current levels with different tap positions. The horizontal scale is generally a multiple of the relay setting which means that a single characteristic can be plotted. However, on this occasion to illustrate the effect the horizontal scale is in amperes.

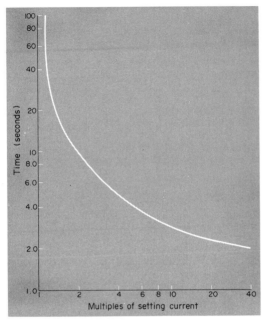

FIG. 5.1   CHARACTERISTIC CURVE. INVERSE DEFINITE–MINIMUM TIME RELAY

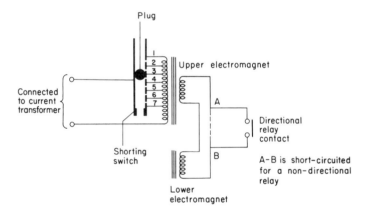

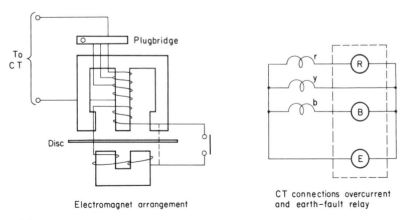

FIG. 5.2

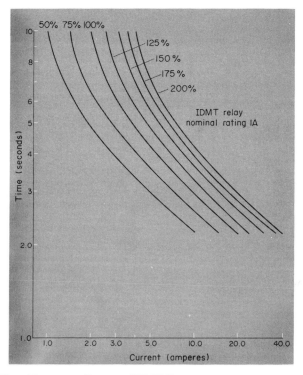

FIG. 5.3   TIME/CURRENT CURVES. IDMT RELAY

As can be seen each tap position has its own curve and the effect of moving the plug is to move the characteristic curve laterally.

It should be noted that should the plug be removed when the relay is in service a switch in the plugbridge automatically connects the 200% tap. This is to prevent open-circuiting of the current transformers.

The time on the curve is the time for the relay disc to move through 180°. By means of the time multiplier, sometimes known as a "torsion head", it is possible to adjust the position of the backstop so that the relay disc is moved towards the contact making position thus decreasing the distance which the disc has to move to make contact. This means that when twice setting current is applied, instead of taking 10 seconds to operate, with only half the distance to travel the relay would operate in 5 seconds. If the distance was 30% of the full travel then the operating time would be 30% × 10 s = 3 s and so on. A setting of 10%, usually designated Time Multiplier Setting (TMS) = 0.1, can be used. Some relays have a 0.05 position marked—this should never be used.

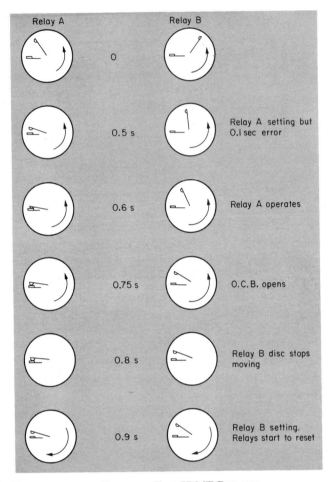

FIG. 5.4   DISCRIMINATION BETWEEN TWO IDMT RELAYS

This type of relay is known as the Inverse Definite Minimum Time (IDMT) relay. Inverse because as the current increases the time decreases—definite minimum because the characteristic appears to approach a definite minimum time. This is not the case, all inverse curves appear to do this but are in fact just as inverse at high values as they are at low values. However, the name has been applied to this class of relay and it is in general use.

The relay has also another peculiarity in that it does not operate at setting current. The disc resetting force is provided by a spring and the definition of setting is when the operating torque is exactly equal to the resetting torque. Unfortunately, the spring torque depends on the disc position, the more it is wound up the higher the torque. To

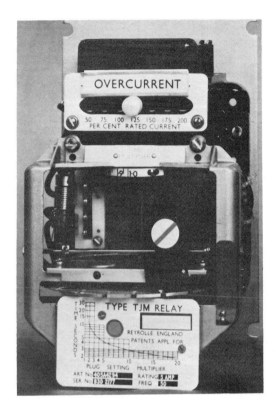

PLATE 5.1   TYPICAL IDMT OVERCURRENT RELAY ELEMENT
(Reyrolle Protection)

compensate for this the disc has some means of reducing the torque produced at low spring wind-up end of the travel either by slots in the disc or by disc shape.

SETTINGS

When determining a setting for an IDMT relay a number of allowances made by BS 142 must be taken into account. BS 142 states that the relay must definitely reset at 70% setting and definitely operate at 130% setting. Modern relays have a reset figure of 90% and an operate figure of 110%. These affect the choice of plug setting in two ways.

Firstly, it is important that, under normal full-load conditions, the relay occupies the fully reset position. To ensure this a setting must be chosen so that the full-load current does not exceed 0.9× setting. This is to ensure definite resetting under the condition where the current reverts to full-load after a through-fault which has caused some relay disc movement.

Secondly, it is preferable that, during temporary moderate overload conditions, there should be no disc movement. This means that a plug setting should be chosen so that the overload current does not exceed 1.1 times the setting. This applies where the overload is of a long duration. Where short-term overload occurs movement of the disc can be permitted providing that the overload does not persist for a time longer than the relay operating time at the overload current. It is not possible to determine relay operating time at this stage and therefore the plug setting should be selected on a basis of normal load current and long-term overload and operation checked under short-term overload conditions at the selected time multiplier setting.

When determining a current setting it is important that the setting be as high as, or higher than, the current setting of the preceeding relay, i.e. the relay which, in the event of a fault involving both relays, will operate first.

If the relay which should operate first was given a current setting higher than the following relay, at lower values of current maldiscrimination may result. Therefore the general rule is that the current setting of a relay nearer the source must always be the same or higher than the setting of the preceding relay.

TIME-MULTIPLIER SETTING

There are four factors which affect the discrimination period between relays.

Two of these are allowances made by BS 142, namely,

1. A variation from the ideal characteristic curve for which an error in time of 0.1 s is used for calculation purposes.
2. Overshoot, i.e. disc movement after the removal of current. Although BS 142 allows 0.1 s the relay performance is much better than this, 0.05 s is used and even this figure gives a large safety margin.

The other two factors are concerned with application.

3. Circuit-breaker operating time. 0.15 s is allowed.
4. Contact gap. To ensure that a relay still has a short distance to travel when the fault is cleared by the relay with which it is discriminating. A time of 0.1 s is allowed.

Assuming that all errors and allowances are additive then the discrimination period should not be less than

$$0.1 + 0.05 + 0.15 + 0.1 = 0.4 \text{ s}.$$

The minimum discrimination period of 0.4 s is the time interval between relay operation at the maximum fault level. If the discrimination period is achieved under these circumstances then at all lower levels the current time interval will be greater.

To illustrate the significance of the discrimination period consider two relays which are associated with two circuit-breakers A and B. Relay A at the maximum fault level will operate in 0.5 s and relay B has been set to operate in $0.5 + 0.4 = 0.9$ s at the same fault level.

When a fault occurs beyond A, both relays respond.

After 0.5 s relay A has operated.

The disc of relay B has moved part way and is 0.4 s from tripping.

Perhaps relay A has a 0.1 s error and does not operate until 0.6 s.

Relay B is now 0.3 s from tripping.

Relay A has energised the trip coil of the circuit-breaker which is opening and clears the fault after $0.6 + 0.15 = 0.75$ s.

Relay B is 0.15 s from tripping.

Although the fault current has disappeared the disc of relay B continues to move because of inertia and it will be 0.05 s before it comes to rest and then starts to reset.

Relay B is 0.1 s from tripping.

As can be seen relay B has discriminated successfully with relay A.

BS 142 also suggests that a 0.1 s allowance should be made because relay B in the above example may be fast. There are a number of reasons why this is not considered necessary.

1. It is unlikely that two relays on one installation will have maximum errors both positive and negative.
2. Discrimination is calculated at maximum fault level which is usually higher than the actual fault level.
3. At high fault levels the current transformers will have some ratio error and will deliver to the relay a current less than calculated value.

APPLICATION

The first relay to be considered is the one furthest removed from the source. This relay is usually set to discriminate with protection on the distribution circuits, e.g. fuses, instantaneous high-set elements and direct-acting overcurrent trips at the maximum, i.e. three-phase, fault level which affects both the relay under consideration and the distribution circuit protection.

If the first relay is to discriminate with fuses, miniature or moulded-case circuit-breakers or direct-acting trips then an operating time at

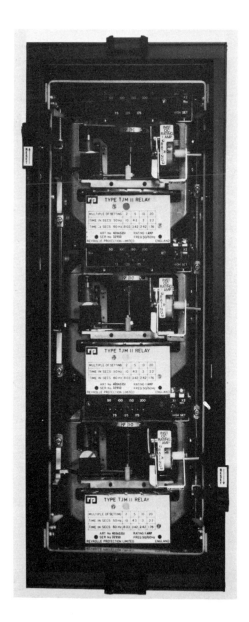

PLATE 5.2　THREE-POLE IDMT OVERCURRENT RELAY WITH INSTANTANEOUS HIGH-SET ELEMENT (Reyrolle Protection)

maximum fault current of 0.3 s would give satisfactory discrimination. This figure is made up of the fuse-operating time or the device-tripping time plus 0.1 s for error, 0.05 s for overshoot and 0.1 s for contact gap.

If the first relay is to discriminate with a device which trips a circuit-breaker then a circuit-breaker time must be added plus, say, 0.1 s for the device-operating time giving a total time of 0.5 s.

To determine a current setting a full statement of the facts pertaining to that relay should be made. For example, a typical statement could be:

> CT ratio 800/5.
> Normal load current 770 A.
> Overload 920 A for 25 s.
> To discriminate with instantaneous high-set relay.
> Maximum fault current which can flow in instantaneous high-set relay is 10,000 A.

1. Consider the normal load condition. The normal load current must not be greater than 90% of the relay setting.

$$\text{Current setting } \frac{770}{0.9 \times 800} \times 100\% = 107\%.$$

2. Consider the long-term overload condition. The overload current must not be greater than 110% of the relay setting.

$$\text{Current setting } \frac{920}{1.1 \times 800} \times 100\% = 105\%.$$

The setting selected would be 125% which is the next highest available relay setting.
The primary current setting is 125% × 800 A = 1000 A.
Fault current as a multiple of setting

$$= \frac{10,000}{1000} = 10.$$

At 10× setting the relay operating time for full travel, $t_1 = 3$ s. Required operating time $t = 0.5$ s. Amount of disc travel to produce operation in 0.5 s is the Time Multiplier setting

$$\text{TMS} = \frac{t}{t_1} = \frac{0.5}{3} = 0.17.$$

Relay setting is therefore 125%, 0.17.

Following the establishment of the setting the performance of the relay is calculated at the fault level just beyond the relay location. A new time is calculated and as the fault level and therefore the multiple of setting is increased this time would be shorter and would be the time which would determine the operating time of the next relay to be considered. It may be that the change in fault level throughout the system under consideration is so small as to make practically no difference to the relay-operating time. Under these circumstances it is permissible to perform all calculations using the fault level at incoming busbar.

Consider the next relay. Once again a full statement of its status should be made; the current setting should be determined and the fault current as a multiple of setting calculated. The operating time for full disc travel should be obtained from the characteristic curve or from the equation $t_1 = 3/\log M$ and the time multiplier setting to give an operating time of 0.4 s plus the recalculated time of the preceding relay determined from

$$t = \text{TMS} \times t_1.$$

DISCRIMINATION WITH FUSES

Discrimination with fuses has already been mentioned and it has been established that the operating time of the first relay can be 0.3 s at the maximum fault level, but it may be that the most onerous condition is at a fault level associated with a fuse-operating time of about 5 s. The best way to check this is by using a template.

Figure 5.5 shows a fuse manufacturer's published curve for 200-A fuse. On this has been plotted the IDMT relay characteristic curve for a setting current of 100 A and a time multiplier setting of 1. If a piece of clear plastic is placed over this and the current line at 100 A, the time line at 1 s and the characteristic curve is drawn on it, then it can be used as a template to represent relay performance at any setting current level and at any time multiplier setting between 0.1 and 1. Figure 5.5 also shows the template in position at a current setting of 1000 A and a time multiplier setting of 0.13. As can be seen the relay will discriminate with a 200 A fuse and will give an operating time of 0.3 s at a maximum fault level of 20,000 A.

If a setting of 800 A had been selected as shown in Fig. 5.6 to achieve discrimination would require a time multiplier setting of 0.43 which would result in an operating time at the maximum fault level of 20,000 A of 0.9 s. So although a setting of 800 A appears to be more attractive the extra time to trip at maximum fault level makes it a very poor choice.

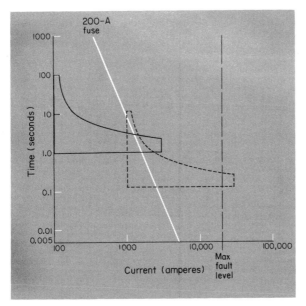

FIG. 5.5   DISCRIMINATION BETWEEN A FUSE AND A RELAY SET AT 1000 A AND A
TIME MULTIPLIER SETTING OF 0.13. TIME AT 20,000 A = 0.3 s

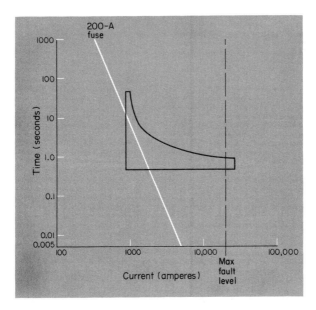

FIG. 5.6   DISCRIMINATION BEETWEEN A FUSE AND A RELAY SET AT 800 A AND A
TIME MULTIPLIER SETTING OF 0.43. TIME AT 20,000 A = 0.9 s

TYPICAL CALCULATION

Figure 5.7 shows a radial distribution system fed at substation A with maximum loads at substations D, C and B of 180 A, 200 A and 220 A, respectively, and fault levels of 6000 A, 7000 A and 8000 A respectively. The largest motor at substation D is 15 kW and takes a current of 6 times full-load current during the starting period of 5 s. Motor full-load current is 28 A. The largest fuses at substation D, C and B are 80 A, 100 A and 125 A, respectively.

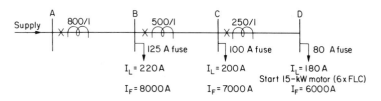

FIG. 5.7 RADIAL DISTRIBUTION SYSTEM

Relay at C

CT ratio 250/1.
Normal max. load current, $I_L = 180$ A.
Max. overload current, $I_T = 180 - 28 + (6 \times 28) = 320$ A,

setting $\dfrac{180}{0.9 \times 250} \times 100\% = 80\%$,

setting $\dfrac{320}{1.1 \times 250} \times 100\% = 116\%$.

Set to 125%,
$I_s = 1.25 \times 250 = 312.5$ A,
$I_f = 6000$ A,

$$M = \frac{6000}{312.5} = 19.2$$

Time for relay full travel, $t_1 = \dfrac{3}{\log 19.2} = 2.32$ s.

Time required to discriminate with a fuse $= 0.3$ s.

Time multiplier setting $= \dfrac{0.3}{2.32} = 0.13.$

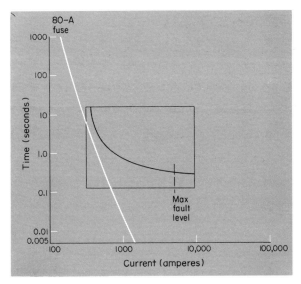

FIG. 5.8   RELAY AT C SET TO 125% = 312 A, TIME MULTIPLIER 0.13 CHECKING
THAT THERE IS DISCRIMINATION WITH AN 80 A FUSE

Check the 80 A fuse characteristic with template (Fig. 5.8)—
satisfactory.

The maximum fault current which the relay at C would have to deal
with is 7000 A for a fault just beyond the CT location (the fault level
is assumed to be the same at this point as it is at the busbar).

Operating time of the relay at C at 7000 A,
i.e. at a multiple if setting of

$$M = \frac{7000}{312.5} = 22.4$$

is TMS $\times t_1 = 0.13 \times \dfrac{3}{\log 22.4}$

$$= 0.13 \times 2.22 = 0.29 \text{ s}.$$

Consider the relay at B.

CT ratio = 500/1,
$I_L = 200 + 180 = 380$ A,
$I_T = 200 + 320 = 520$ A,

setting $\dfrac{380}{0.9 \times 500}$ 100% = 84%,

setting $\dfrac{520}{1.1 \times 500}$ $100\% = 95\%$,

set to 100%,

$I_s = 500$ A,

$I_f = 7000$ A,

$$M = \frac{7000}{500} = 14,$$

$$t_i = \frac{3}{\log 14} = 2.6 \text{ s},$$

$t = 0.4 + 0.29 = 0.69$ s,

$$\text{TMS} = \frac{0.69}{2.6} = 0.27.$$

Check, by template, that there is discrimination with a 100 A fuse. Operating time of relay at B at 8000 A.

$$M = \frac{8000}{500} = 16,$$

$$t = 0.27 \times \frac{3}{\log 16} = 0.67 \text{ s}.$$

Consider the relay at A

CT ratio 800/1,

$I_L = 380 + 220 = 600$ A,

$I_T = 520 + 220 = 740$ A,

setting $\dfrac{600}{0.9 \times 800} \times 100\% = 83\%$,

setting $\dfrac{740}{1.1 \times 800} \times 100\% = 84\%$,

set to 100%,

$I_s = 800$ A,

$I_f = 8000$ A,

$M = 10$,

$t_1 = 3$ s,

$t = 0.67 + 0.4 = 1.07$,

$$\text{TMS} = \frac{1.07}{3} = 0.36.$$

Check, by template, that there is discrimination with a 125 A fuse.

Summary

Relay	CT ratio	Plug setting	TMS
C	250/1	125%	0.13
B	500/1	100%	0.27
A	800/1	100%	0.36

Radially fed power systems have a very low integrity, as a single fault can result in the loss of supply to a number of substations, and consequently the system is not generally used. The more usual method of supply would be by a further interconnection betwen the remote substation and the supply substation as shown in Fig. 5.9.

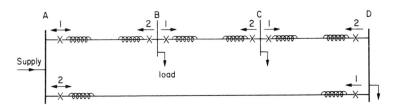

FIG. 5.9   RING DISTRIBUTION SYSTEM

Unfortunately this means that a simple time-graded overcurrent system cannot be used as power and fault current can flow in both directions. To provide discrimination directional relays must be used. These are arranged to operate when power and fault current flow into a feeder and to restrain when power and fault current flow out of the feeder. The contacts of the relay are arranged in conjunction with IDMT overcurrent relay to allow measurement when closed and prevent measurement when open.

At the supply substation there is no necessity to use directional relays as the only possible way that the power and fault current can flow is into the feeder.

The arrows on the system diagram indicate the direction of current flow which will operate the relay at that location. Examination of the system will show that, for protection purposes, the system can be regarded as two radial systems. One concerned with relays A1, B1, C1 and D1 and the other involving relays A2, D2, C2 and B2.

The fault level at the end of each feeder would be calculated by assuming the circuit-breaker at A2 is open when calculating the fault level for relays A1, B1, etc., and the circuit-breaker at A1 is open when calculating the fault level for relays A2, B2, etc. When the circuit is fully connected the fault flow in each direction will be less than that calculated with either circuit-breaker A1 or A2 open and therefore discrimination will be assured.

In many installations of this type it would be usual for all the circuit-breakers and feeders to have the same rating and all the current transformers would have the same ratio. It may also be difficult to determine the actual load current demand at each sub-station bearing in mind that protection settings are required before the system is energised. Under these circumstances a simplified approach can be made which may be modified later in the light of experience.

Figure 5.9 shows the system diagram. The circuit-breakers and feeders are rated at 800 A, the CT ratio at each circuit-breaker is 800/1.

The fault level at the incoming busbar is, say, 12,000 A.

Set all relays to 100% = 800 A.

$$\text{Multiple of setting} = \frac{12,000}{8} = 15,$$

$$t_1 = \frac{3}{\log 15} = 2.55 \text{ s}.$$

Relays D1 and B2

$$t = 0.5 \text{ s},$$

$$\text{TMS} = \frac{0.5}{2.55} = 0.2.$$

Relays C1 and C2

$$\text{TMS} = \frac{0.9}{2.55} = 0.35.$$

Relays B1 and D2

$$\text{TMS} = \frac{1.3}{2.55} = 0.51.$$

Relays A1 and A2

$$\text{TMS} = \frac{1.7}{2.55} = 0.67.$$

Summary

Relay	CT ratio	Plug setting	TMS
A1	800/1	100%	0.67
A2	800/1	100%	0.67
B1	800/1	100%	0.51
B2	800/1	100%	0.2
C1	800/1	100%	0.35
C2	800/1	100%	0.35
D1	800/1	100%	0.2
D2	800/1	100%	0.51

After settings have been calculated it is a good plan to prepare a diagram showing the characteristics of each relay so that the performance of the whole scheme can be seen at a glance. Figure 5.10 shows such a diagram for the radial system.

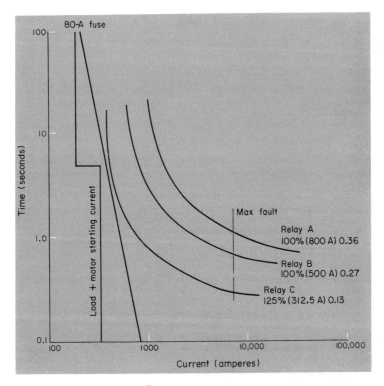

FIG. 5.10   DISCRIMINATION CURVES

The curves are generally plotted on log-log paper covering the time range of all the protection and a current range from minimum load current to maximum fault current. Here again, a template of the characteristic curve is useful. On this occasion the shape of the curve can be cut out so that the characteristic curve can be drawn when the template is in position.

Fuse characteristics, instantaneous relay characteristics, etc., can also be plotted so that an overall picture is shown.

EARTH-FAULT PROTECTION

Earth-fault relays should be used where there is a possibility that the earth-fault current will be limited, e.g. a resistance earthed system or in a low voltage system where even a low impedance can substantially reduce the earth-fault level.

The usual practice is to have two overcurrent relays, one in the red and one in the blue phase with the third element as an earth-fault relay.

The relay is identical to the overcurrent relay except that the primary winding has more turns. This gives a range of settings to, say, 20% to 80% or, less usually, 10% to 40% of rated current.

Where time grading and phase fault stability is required BS 3938 advocates the use of class P5 current transformers in which the product of rated burden and rated accuracy limit factor approaches 150. These are declared to be suitable for ensuring phase-fault stability up to 10 times the rated primary current and for maintaining time grading of earth-fault relays up to 10 times the relay setting, providing that the rated accuracy-limit factor is not less than 10, the burden of the relay at setting does not exceed 4 VA, the phase-fault burden is not greater than 50% of the rated burden and the relay is set at not less than 30%. In fact there are a number of factors which alleviate the situation the main one being that the impedance of the relay coil decreases as the current increases owing to saturation.

The product of rated burden and accuracy limit factor is virtually the knee-point voltage and the recommendation above is based, for a 1 A CT, on 150 V, i.e. a burden of 15 VA and an accuracy limit factor of 10.

The maximum phase-fault relay burden would be 7.5 VA, i.e. 7.5, and with an earth-fault relay having a burden of 4 VA at setting and a setting of 0.3 A its impedance would be

$$\frac{4}{0.3^2} = 44.4 \ \Omega.$$

So the total impedance in circuit would be 44.4 + 7.5 = 51.9 which would give a current of 150/51.9 = 2.9 A, i.e. approximately 10 times the earth-fault relay setting.

When the relay is subjected to 10 times its setting current its impedance is reduced to less than half that at setting because of saturation. This means that the knee-point voltage could be some-what less than 150 V and accurate time grading still be maintained. In fact a setting of less than 30% can easily be tolerated and accuracy maintained beyond the 10 times setting with a current transformer having a knee-point voltage of 150 V. A 20% setting is probably the best that can be used.

In most cases it is the earth-fault relay that determines the CT requirements. The usual IDMT relay has a burden of 3 VA at setting so that a relay set at 20% of 1 A will have an impedance of setting of

$$\frac{3}{0.2^2} = 75 \, \Omega$$

which would decrease to 30 Ω at 10 times setting. So the voltage to drive 2 A through the relay would be 2 × 30 = 60 V. If a relay with a 10% setting was chosen the impedance would be 300 Ω which at 10 times setting would be 120 Ω and therefore 120 V would be required to drive 1 A through the relay. This means that larger current transformers would be required. In fact a relay with a 10% setting is not twice as sensitive as a relay with a 20% setting as the overall setting depends on the sum of the relay current and the magnetising current of the three current transformers at the setting voltage, e.g.

20% relay.

$$\text{Voltage at setting} = \frac{3}{0.2} = 15 \, \text{V}.$$

Say CT magnetising current at 15 V = 0.01 A.
Overall setting = 0.2 + 3 × 0.01 = 0.23 A.

10% relay.

$$\text{Voltage at setting} = \frac{3}{0.1} = 30 \, \text{V}.$$

Say CT magnetising current at 30 V = 0.02 A.
Overall setting = 0.1 + 3 × 0.02 = 0.16 A.

If the magnetising current was high enough the 10% relay would have an overall setting higher than the 20% relay but with modern materials this is very unlikely. A relay with a 20% setting seems to provide optimum setting unless there is a restriction imposed by a limited CT knee-point voltage. In this case a higher setting would have to be chosen to reduce the impedance of the coil.

Care should be taken to ensure that under minimum conditions the relay will receive at least 2× setting current. This is particularly important considering ring systems as the earth-fault current is fed from both ends and one relay will receive less than half the available current. Having determined the plug setting, the time multiplier setting is calculated in the same manner as for phase-fault relays using the maximum earth-fault current. For stability under phase-fault conditions a time multiplier setting of less than 0.1 s should not be chosen.

VERY INVERSE CHARACTERISTIC

One of the drawbacks of a protection system using IDMT relays is that the nearer the fault is to the source of power the slower the overall fault clearance time.

The situation can be improved if there is a large difference in fault level at various parts of the system by the use of relays with a very inverse characteristic. This characteristic is shown in Fig. 5.11 and its equation is

$$t = 1.6(\log M)^{-2}.$$

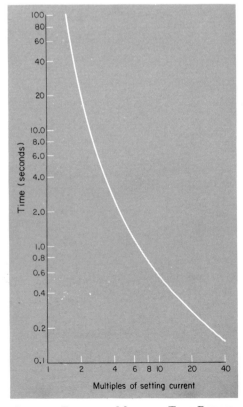

FIG. 5.11 VERY INVERSE DEFINITE–MINIMUM TIME RELAY

EXTREMELY INVERSE CHARACTERISTIC

It will already have become apparent that the normal IDMT relay has a characteristic which is very different from that of a fuse. For relays which are to discriminate with fuses a much closer agreement with the fuse characteristic is provided by relays with an extremely inverse characteristic. This characteristic is shown in Fig. 5.12 and its equation is

$$t = 0.6(\log M)^{-3}.$$

The large change in fault level and the use of fuses means that the most likely application of these two types of relay will be on low-voltage systems.

HIGH MULTIPLES OF SETTING

In some cases, particularly where low ratio current transformers are used, the calculated multiple of relay setting may be very high. There is a limit to the amount of current that the relay will withstand

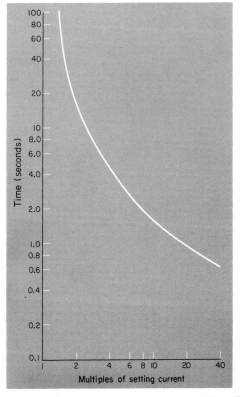

FIG. 5.12   EXTREMELY INVERSE DEFINITE–MINIMUM TIME RELAY

but it is unlikely that this limit would be reached as the current transformer would probably saturate before this point.

The amount of current which would flow in the relay is somewhat indeterminate and so in order to calculate a time setting an arbitrary value must be assumed.

If a maximum value of 30 times setting for overcurrent relays and 20 times setting for earth-fault relays a reasonably accurate result will be obtained.

Chapter 6

Unit Protection

Protection schemes which operate on the principle of discrimination by comparison are known as unit schemes. This is because they protect only the unit with which they are associated and do not provide the back-up protection which all discrimination by time schemes provide.

Most unit schemes are based on the Merz–Price principle which basically is that if the current flowing into the protected unit is the same as the current leaving then the fault is not in the protected unit and the protection should not trip. If there is a difference in either phase or magnitude between input and output then the fault is in the unit and the protection should trip.

The schemes depends on the relay being connected to the centre point of a balanced system. Figure 6.1 shows the connections. CT1 and CT2 are identical—having not only the same turns ratio but also

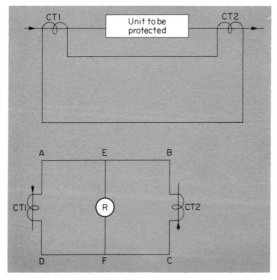

FIG. 6.1   MERZ–PRICE PROTECTION

103

the same magnetising characteristic and resistance. In addition the two connecting leads between the current transformers are identical.

Therefore, when the same current flows in the primary of both current transformers, the secondary current in each CT secondary winding will be the same. The voltage across each CT will be the same and will be determined by the product of the current and the resistance of one connecting lead. In other words all the voltage produced across CT1 will be dropped across lead A–B and all the voltage across CT2 will be dropped across C–D. This means that A and C are at the same voltage as are B and D, and E and F which are half-way along each connecting lead. Therefore if a relay is connected between A and C or B and D or E and F then, under the conditions where both current transformers carry the same current, the relay current will be zero.

When the current in the two current transformers is not the same the difference between the CT secondary currents is passed through the relay.

There are certain difficulties which arise if the scheme is used in this form because it is impossible to produce identical current transformers. There will always be differences, however slight, in the magnetising characteristic and this leads to instability in the scheme during through fault conditions.

Figure 6.2 shows the magnetising characteristics of two current

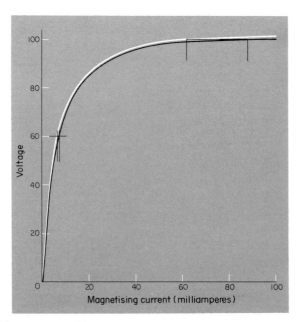

FIG. 6.2 MAGNETISING CHARACTERISTICS OF TWO SIMILAR CURRENT TRANSFORMERS USED IN A MERZ–PRICE SYSTEM

transformers. As can be seen they are fairly closely matched. At 60 V there is a low magnetising current and the difference between the two magnetising currents, which is known as spill current, would flow in the relay. This would not present a problem as the current would be much less than the relay setting. At 100 V the current transformers are saturated, the magnetising current is greatly increased as is the difference in magnetising current. The spill current would exceed the relay setting and cause operation. The solution would seem to be to ensure that the CT knee-point voltage is not exceeded. However, this is not possible because of the behaviour of the power system in the fraction of a second following the instant at which a fault occurs. This requires some explanation.

When a resistance is switched on to an a.c. supply an a.c. current will flow and its wave-form will be sinusoidal as shown in Fig. 6.3(a).

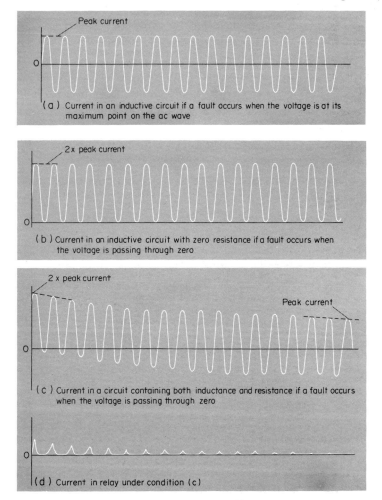

(a) Current in an inductive circuit if a fault occurs when the voltage is at its maximum point on the ac wave

(b) Current in an inductive circuit with zero resistance if a fault occurs when the voltage is passing through zero

(c) Current in a circuit containing both inductance and resistance if a fault occurs when the voltage is passing through zero

(d) Current in relay under condition (c)

FIG. 6.3   CURRENT WAVE-FORMS UNDER FAULT CONDITIONS

When a pure inductance is switched on to an a.c. supply at the instant when the rate of change of voltage is zero, i.e. at maximum voltage, then the current will be the same as if a resistance has been switched in.

If the inductance was switched when the voltage was zero, i.e. at maximum rate of change, a current will flow which does not reverse. Figure 6.3(b) shows the wave-form which is as though a d.c. current, of the same value as the peak current, has been added to the a.c. current.

Although a normal power system is very inductive it has some resistance and so, although the wave-form would start as in Fig. 6.3(b), it would quickly decay to the wave-form shown in Fig. 6.3(a). The time taken to die away would depend on the ratio of the system inductance and resistance. A typical wave-form is shown in Fig. 6.3(c).

The deviation from the normal sinusoidal wave-form is known as a transient and is virtually a d.c. component which decays exponentially. That is to say that if it reached half the initial value in a certain time after another period of the same time it will have halved again and so on.

The effect of the d.c. component on the current transformer circuit is to push it well beyond the knee-point flux into saturation which results in a large spill current and relay operation. Figure 6.3(d) shows a typical wave-form of the spill current.

The possibility of a fault occurring at or near a voltage zero is high on a three-phase system as there is a voltage zero 300 times every second and there is always some proportion of the transient in at least two phases.

One solution to the problem of this spill current would be to introduce a time delay of, say, 0.5 s so that the transient would have disappeared before the relay was called on to operate but there is a very simple alternative which is in general use—the high-impedance relay.

This is a voltage-operated relay but, although it is connected in the CT circuit in exactly the same manner as the current-operated relay in the Merz–Price system, the principle of operation is quite different.

CT1 is regarded as a perfect current transformer which will maintain accurately its ratio up to the maximum fault level—the secondary current will be I_f. It is not necessary to connect the relay across equipotential points nor is it necessary to balance the leads between the relay and current transformers. Let the lead resistance be R_L. Under the worst possible circumstances if CT2 gave no output whatsoever the maximum voltage which would appear across the relay would be:

PLATE 6.1   HIGH-IMPEDANCE RELAY WITH VOLTAGE SETTING LINKS
(Reyrolle Protection)

$$V = I_f(R_L + R_{CT})$$

where R_{CT} is the resistance of the secondary winding of CT2.

Therefore, if the voltage setting of the relay is higher than V the relay cannot operate under through-fault conditions even if there was no output from CT2.

In fact under through-fault conditions the output from CT2 will be similar to that of CT1 and even under the transient condition the voltage across the relay will be only a fraction of its setting and so stability to through faults is ensured.

Under internal fault condition the current in one of the current transformers will be absent or reversed and a voltage in excess of relay setting will be produced across the relay. Actually such a high voltage would be produced that it is necessary to limit its value by means of a non-linear resistor.

Consider a practical case:

let $I_f = 30$ A. $R_L = 0.8\,\Omega$, $R_{CT} = 3.2\,\Omega$,
$V = 30(0.8 + 3.2) = 120$.

The relay has a setting current of, say, 0.02 A.

Therefore the total relay circuit resistance is

$$R = \frac{120}{0.02} = 6000\,\Omega.$$

During an internal fault when the full output of CT1, i.e. 30 A, would be driven through the relay a voltage of

$30 \times 6000 = 180$ kV would result.

In fact it would be a lot less than this because a large proportion would be used as CT magnetising current. Even so, the situation is completely unacceptable and necessitates the use of a non-linear resistor.

A non-linear resistor is a device which has a characteristic such that if the voltage across it is doubled then the current, instead of doubling as would be the case with a normal resistor, increases by 32 times, e.g.

Voltage across relay	Current in non-linear resistor
120	0.01 A
240	0.32 A
480	10.24 A
600	30.00 A

As can be seen the 30 A would pass through the non-linear resistor and the voltage would be limited to 600 V. The unit would be about 150 mm dia., as it has to be capable of dissipating something like $600\,\text{V} \times 30\,\text{A} \times 3\,\text{s} = 54{,}000$ j.

RELAYS

The relays used are generally attracted-armature types and sometimes incorporate a tuned circuit to decrease the sensitivity to the d.c. components of the input waveforms. The relay usually has a setting of 0.02 A and suitable resistance units to give a range of settings from 15 or 25 volts up to about 200 V. The resistor may be an external unit or, more usually, mounted inside the relay case in which case adjustment is by rheostat, shorting links or plugbridge. The rheostat would have a dial calibrated in setting voltage, the shorting links enable resistance values equivalent to 5, 10, 20, 40 and 80 V to be connected in or short-circuited to allow any voltage from 15 V, the coil voltage, to

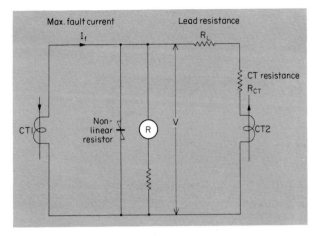

F<small>IG</small>. 6.4 H<small>IGH-IMPEDANCE</small> R<small>ELAY</small>

180 V to be selected. If, for example, a setting of 70 V is required, the 70 V would be made up of coil voltage and the 40, 10 and 5 V units as shown in Fig. 6.5.

The plugbridge would have resistance units between the plug positions. These could be ordinary resistors or could be non-linear resistors. The advantage of the latter is that at twice setting voltage up to 6 times setting current will flow in the relay thus producing a high-speed operation when the setting voltage is exceeded.

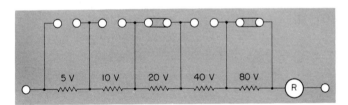

F<small>IG</small>. 6.5 H<small>IGH-IMPEDANCE</small> R<small>ELAY</small> S<small>ET TO</small> 70 V

APPLICATION

The high-impedance relay is used for differential protection of generators, motors, busbars, reactors, balanced and restricted earth-fault protection of transformers. Although its application is wide the method of determining setting is the same in all cases.

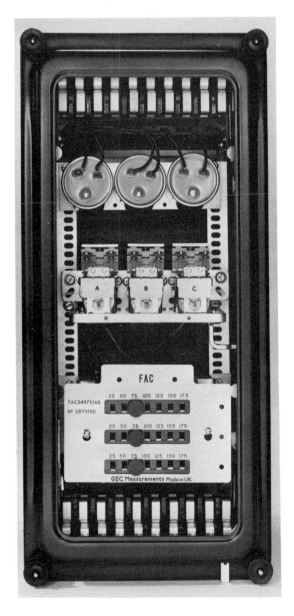

PLATE 6.2   THREE-POLE HIGH-IMPEDANCE RELAY. Voltage setting selected by Plugbridge (GEC Measurements)

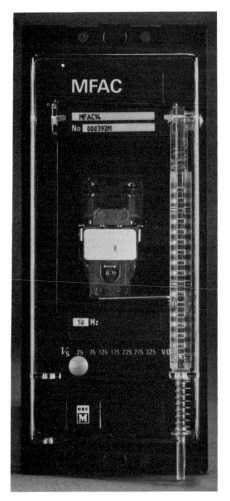

PLATE 6.3   A MICROPROCESSOR-BASED HIGH-IMPEDANCE RELAY
(GEC Measurements)

The basis is the equation

$$V = I_f(R_L + R_{CT}),$$

I_f the maximum fault current which can flow to a through fault or, in the case of earth fault relays, the maximum phase-fault current which can flow to an internal or through fault. R_L is the resistance of the leads in a CT circuit between the current transformer and the point where the current transformers are connected in parallel. The highest value is taken for the purpose of calculation as this gives the worst case. In Fig. 6.6, 0.7 Ω would be the figure used. The value of R_L should include the impedance of any devices which are connected in

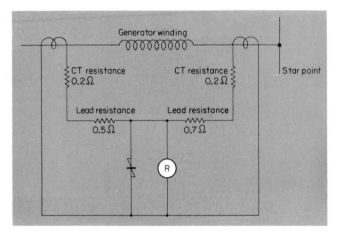

FIG. 6.6 SINGLE-PHASE DIAGRAM OF A HIGH-IMPEDANCE RELAY USED FOR GENERATOR PROTECTION

the CT circuit, e.g. overcurrent relays. R_{CT} is the resistance of a current transformer secondary winding. All the current transformers in a group will have similar values a possible exception being in restricted earth-fault protection where the neutral current transformer may differ and in all probability will also have the longest leads.

The values of resistance for the current transformers and the leads will usually be at 20°C, whereas they could possibly operate at around 70°C which means that the value of resistance could be 1.2 times greater. This should be taken into account.

Having determined the maximum fault current and the lead and CT resistance it is a simple matter to calculate the setting voltage, which is

$$V = I_f(R_L + R_{CT})$$

in the case of relays which have a tuned circuit or 1.5 times this value otherwise. In cases where there is some uncertainty the higher setting should be used.

The relay type will determine the actual setting applied. If the relay has an external resistor or a calibrated rheostat then the calculated value would be the setting, in other cases the next highest setting would be used.

CT KNEE-POINT VOLTAGE

A knee-point voltage of at least twice the relay setting voltage is required to ensure that the relay will operate with some speed which would not be the case if the knee-point voltage was only just in excess of the setting voltage.

OVERALL SETTING

One of the great advantages of a unit protection scheme is that the current setting can be much less than the normal load current. This is because the relay operates on the difference between the current flowing into the unit and the current flowing out which, under healthy conditions, is zero.

The current setting of the relay is typically 0.02 A but the overall setting is greater than this as the current taken by the non-linear resistor and the magnetising current taken by the current transformers at setting voltages have to be added. For example, say the magnetising current of each CT is 0.012 A at a setting voltage of 60 V. The current taken by the non-linear resistor is negligible at this voltage and so the overall setting is

$$I_S = 0.02 + 2 \times 0.012 = 0.044 \text{ A}$$

or if the CT ratio is 500/1 a primary setting of

$$500 \times 0.044 = 22 \text{ A}.$$

In some schemes there are more than two current transformers involved—three or four in the case of earth-fault protection and many more in busbar-protection schemes.

It is sometimes felt necessary to desensitise a unit protection scheme. This can be achieved by connecting a resistor across the current transformers in order to increase the current at setting voltage.

THE RESIDUAL CONNECTION

When current transformers are connected as shown in Fig. 6.7 they are said to be residually connected. This is the normal method of connection for the detection of earth-faults and is, of course, a Merz–Price system.

If the relays are IDMT relays then any unbalance of the CT secondary current during the transient period will have disappeared before any relay operation could take place. If the residually connected relay is an attracted armature or any instantaneous relay then it must be converted to a voltage relay by connecting a resistor in series with its coil. The resistor is known as a stabilising resistor and

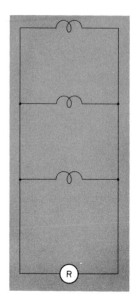

FIG. 6.7   RESIDUALLY CONNECTED CURRENT TRANSFORMERS

its value depends on the calculated setting voltage and the setting current of the relay.

One of the problems which occurs frequently is that the earth-fault element in a motor-protection relay operates during motor starting. Suppose the relay has a 1 A rating; the starting current is 8 times the CT rating and each CT has a resistance of 0.5 Ω and leads of 0.1 Ω. The motor-protection relay overcurrent elements have an impedance of 6.5 Ω which reduces to 3 Ω at 8× rating.

$V = 8(0.5 + 0.1 + 3) = 28.8$ V.

The earth-fault element has a 10% setting and so the total coil circuit resistance including stabilising should be

$$R = \frac{28.8}{0.1} = 288 \ \Omega.$$

From this should be subtracted the coil resistance to give the value of the stabilising resistor. It will be noted that the factors of 1.2 and 1.5 have not been used to allow for temperature rise and lack of coil tuning respectively. These allowances should be made if possible but the overriding constraint is the CT knee-point voltage which must be at least twice the relay setting. In all probability this will limit the voltage to which the earth-fault element can be set to less than the calculated value.

BUSBAR PROTECTION

The high-impedance relay is eminently suitable for the protection of busbars as speed and stability are the prime requirements. Speed to limit damage and stability to limit disruption.

On many industrial installations, busbar zone protection is not considered necessary or its use is limited to important substations. This is justified by the fact that a busbar fault is very unlikely and even if one should occur it would be cleared by other protection. If the fault is cleared by other protection, however, there would be a much wider disruption, possibly leading to loss of supply to a large area, than if they were cleared by busbar protection. This is because with busbar protection the installation would be divided into discrete zones and therefore only the faulty zone would be disconnected. In the early days there was some reluctance because of the fear of inadvertent tripping with the consequent loss of supply. The possibility of this occurring with a modern high-stability system is extremely remote.

Most busbar faults involve earth and with phase-segregated switchgear this is the only type of fault. However, if there is a possibility of faults which do not involve earth, protection must be provided to cater for this condition. Practically all schemes depend on the operation of two protection systems which check that there is a busbar fault and range from simple earth-leakage detection to the comprehensive fully discriminative schemes with overall check features.

Busbar protection must detect and clear faults within the busbar zone and not trip for a fault at any other location. If current transformers, all having the same ratio, are mounted on every circuit connected to the busbar and are connected in parallel, as shown in Fig. 6.8, then a fault external to the busbar will result in a balance of all CT currents and the high-impedance relay will not operate. If the fault is on the busbar then the protection will operate.

The busbar shown in Fig. 6.8 can be divided into two parts by means of the bus-section switch and so, if current transformers are installed on each side of the bus-section switch, the protection can also be divided into two parts, as shown in Fig. 6.9. This means that for a busbar fault on either side of the bus-section switch only the faulty section need be tripped leaving the healthy section intact.

It should be noted that in the bus-section switch current transformers are cross-connected, i.e. those in the A zone are connected to the B zone protection and vice versa. This ensures that all the busbar, including the bus-section switch, is protected by this overlapping. A fault on the bus-section switch itself will result in a

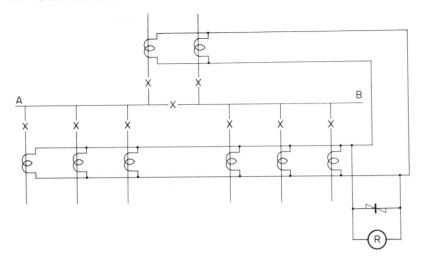

FIG. 6.8   BUSBAR PROTECTION. OVERALL PROTECTION—ALL CT SECONDARY
WINDINGS CONNECTED IN PARALLEL

complete loss of the busbar. In some cases, where a high-integrity
supply is required a double bus-section arrangement is used as shown
in Fig. 6.10. In this case the busbars between the two bus-section
switches are treated as a separate zone.

It has been mentioned that with phase-segregated switchgear there
is no possibility of a phase fault and therefore any fault must be to
earth. Under these circumstances each set of three current transformers on each circuit can be residually connected which means that

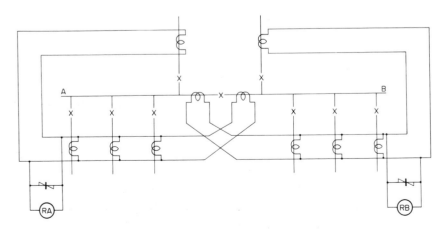

FIG. 6.9   BUSBAR PROTECTION. WITH 2 ZONES

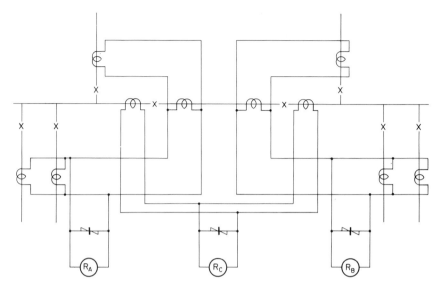

FIG. 6.10 BUSBAR PROTECTION. DOUBLE BUS SECTION

every CT in a zone will be connected in parallel with a single-element high-impedance relay connected in parallel also.

One of the principles used in the protection of busbar is that tripping should not be caused by the operation of one relay and in view of this various methods of checking that there actually is a fault are used. There are three methods of providing a check feature for an earth-fault scheme.

1. Fully discriminative check.
2. Frame-leakage check.
3. Neutral check.

1. Fully Discriminative Check

This is simply another fully discriminative scheme using another set of current transformers in each circuit connected to the busbar and another high-impedance relay. It is an overall check system and therefore there is no need to provide extra current transformers in the bus-section switch or switches. Figure 6.11 shows the arrangement and Fig. 6.12 shows the tripping circuit. In the tripping circuit a check tripping relay is used so that tripping does not rely on one tripping relay. Two contacts are in series with each circuit-breaker trip coil and the bus-section switch is tripped by the protection on both zones.

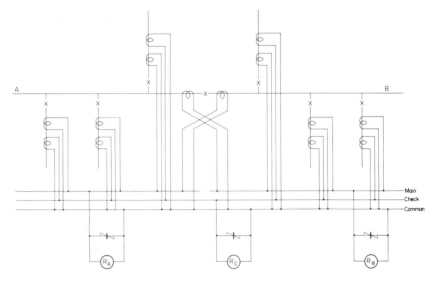

FIG. 6.11   BUSBAR PROTECTION. FULLY DISCRIMINATIVE MAIN AND CHECK
SCHEMES

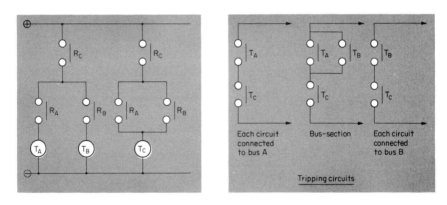

FIG. 6.12   BUSBAR PROTECTION TRIPPING CIRCUITS

2. Frame-leakage Check

In the frame-leakage scheme the switchgear frame must be insulated from earth. The insulation need not be elaborate, it is sufficient to see that there is no contact between the switchgear and any earthed metalwork such as concrete reinforcing bars. The cable glands also need to be insulated from the switchgear, the sheaths and armouring being connected to the station earth bar. The switchgear is also con-

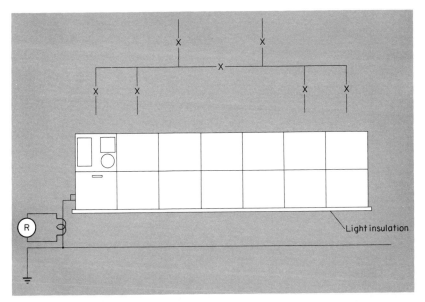

FIG. 6.13   FRAME LEAKAGE CHECK SCHEME

nected to the earth bar but through a current transformer to which is connected an instantaneous earth fault relay which will detect any current flow from the switchgear frame to earth. Figure 6.13 shows the arrangement.

Where there is a typical separation between the switchgear of each bus-section the frame leakage scheme can be used as a main protection scheme as shown in Fig. 6.14. In this case the bus-section switch is treated as a separate zone but is tripped for a fault on either bus-section. The check is provided by a neutral check scheme. An alternative scheme is to include the bus-section switch as part of one of the bus-section zones and to arrange the tripping circuit to cater for the condition where the fault is on the bus-section switch by a discrimination by time arrangement. If the fault is anywhere but the bus-section switch then that section will trip immediately. If the fault is on the bus-section switch, section A is tripped immediately and, as the fault will persist, after a time-delay section B will trip. It is difficult to see what advantage this scheme has over the previous scheme as any saving in cost would be marginal.

In some cases a frame-leakage scheme is the only protection and therefore inadvertent operation of one relay could result in tripping. The provision of a short time-delay relay will remove the possibility of tripping owing to operation of the relay by mechanical shock.

3. Neutral Check

In this scheme the incoming supplies are monitored for earth faults

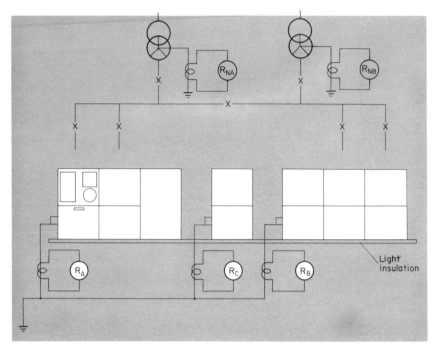

FIG. 6.14   FRAME LEAKAGE BUSBAR PROTECTION WITH NEUTRAL CHECK

by current transformers in the neutrals of the incoming supply transformers if these are located at the substation (Fig. 6.14) or if not, by residually connected current transformers on each incoming feeder. The relay will, of course, be unrestricted which means that it will operate for any system earth fault but as the main protection will be a fully discriminative or a frame-leakage scheme which will only operate for a busbar fault, the situation is satisfactory.

A variation of the residually connected current transformers is the use of core balance current transformers on each incoming supply. In this case the cable sheath earth must by-pass the current transformer.

PHASE AND EARTH-FAULT SCHEMES

Where there is a possibility of phase faults the protection must be capable of detecting and clearing them. The protection is usually fully discriminative as in the earth-fault scheme but comparison of current flow is on a phase by phase basis.

The check system is also likely to be fully discriminative but, as an alternative, overcurrent relays on each incoming supply can be used. A simplified diagram of a phase and earth-fault scheme with overall discriminative check is shown in Fig. 6.15.

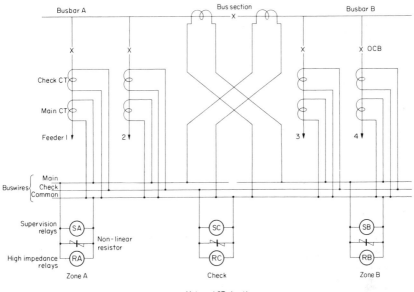

Main and CT circuits

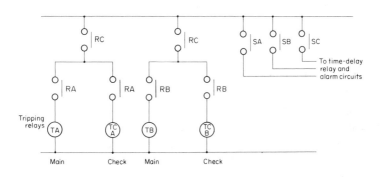

Tripping relay circuits

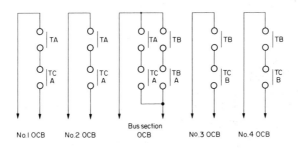

Tripping circuits

FIG. 6.15   SIMPLIFIED BUSBAR PROTECTION

SETTINGS

The setting of the relays is in accordance with the normal high-impedance relay theory, i.e. $V_s > I_f(R_{CT} + R_L)$. The fault current, I_f, is usually taken to be the CT secondary current associated with the rated symmetrical breaking current of the switchgear. This figure is used even for earth-fault schemes where the earth-fault current may be restricted by resistance earthing as stability is required to through-phase faults as well as through earth-faults.

$R_{CT} + R_L$ is the resistance of a current transformer plus the loop resistance of the connection between it and the relay. The value used in the equation is the highest which, as all the current transformers will probably have similar resistances, is the resistance of the longest connection in the circuit.

From the voltage calculated the actual relay setting is determined. If the relay is set by internal rheostat or external resistor then the calculated value can be used. If the relay setting is adjustable by plugbridge or shorting links then the next highest setting is used.

It should be noted that it has been assumed that the relay is of the type which has a tuned circuit. If this is not the case then the basic equation becomes

$$V_S \quad 1.5 I_f(R_{CT} + R_L),$$

i.e. the setting voltage will be based on a figure 1.5 times that calculated previously.

The knee-point voltage of every current transformer must be at least twice the voltage setting of the relay to ensure that it operates at a reasonable speed under internal fault conditions. The required knee-point voltage will, of course, have been determined in the design stage and therefore some estimation of the value R_L would have been made and specified to the CT designer. At this stage the value of R_{CT} is unknown and the specification would quote the equation

$$V_{KP} \quad 2I_f(R_{CT} + R_L)$$

and give values for I_f and R_L. The value of R_{CT} would be determined by the CT designer. Alternatively an estimate of the CT resistance could also be made and a knee-point voltage quoted with a proviso that a certain CT resistance must not be exceeded.

CURRENT SETTING

From the magnetising characteristic of the current transformers the magnetising current at setting voltage can be obtained. From this the overall current setting of the protection is calculated by summating the magnetising currents of all the current transformers and adding the relay current.

For example, if the setting is 100 V and the magnetising current at 100 V is 0.01 A; there are eight circuits, the relay current at setting is 0.02 A and the CT ratio 400/1. Then for a phase and earth-fault scheme there are eight current transformers in parallel and therefore the overall setting is

$$8 \times 0.01 + 0.02 = 0.1 \text{ A or } 10\%.$$

For an earth-fault scheme twenty-four current transformers would be connected in parallel and the setting would be

$$24 \times 0.01 + 0.02 = 0.26 \text{ A or } 26\%.$$

It is a mistake to use an unnecessarily low setting for a busbar-protection scheme because of the consequences of mal-operation. If the power system is solidly earthed then there will be no shortage of fault current during a busbar fault and a setting equal to the current transformer rating would be satisfactory.

Settings are increased by the connection of a resistance in parallel with the relay circuit which is, of course, in parallel with all the current transformers. The effect of this is to shunt sufficient current to increase the setting to 1 A in the examples given.

Setting voltage 100 V.
Relay + CT magnetising current = 0.1 A.
Therefore resistor current = $1 - 0.1 = 0.9$ A

and the resistance $= \dfrac{100 \text{ V}}{0.9 \text{ A}} = 111 \ \Omega.$

Or for the earth-fault scheme:
Relay + CT magnetising current = 0.25 A.
Resistor current = $1 - 0.25 = 0.75$ A

and the resistor $\dfrac{100}{0.75} = 133 \ \Omega.$

If the power system was resistance earthed and the earth-fault current limited to, say, the CT primary rating—400 A in the example—then a setting of about 100 A or 25% would be suitable.

In the case of the phase and earth-fault scheme the value of the shunt resistor would be

$$\frac{100}{0.25 - 0.1} = 667 \ \Omega$$

For the earth-fault scheme a resistor would not be required.

Because busbar protection requires a high degree of reliability and because, in a fully discriminative scheme, reliability depends heavily on the integrity of the CT circuits supervision schemes are sometimes included to monitor these. The scheme consists of voltage relays connected in parallel with each high-impedance relay but responding to a much lower voltage. A setting of 10 V is usual. The scheme will detect any open circuit CT lead at very low levels of current but will not respond to short-circuited connections. It is assumed that the latter fault is extremely remote.

There are two busbar arrangements which are associated with transmission substations which need to be mentioned. These are the duplicate-busbar substation and the mesh-connected substation.

DUPLICATE-BUSBAR PROTECTION

This would have a normal phase and earth-fault protection scheme. Fully discriminative with a relay for each zone and an overall discriminative check. The unusual feature is that the current transformers in the feeders need to be switched by isolator auxilliary contacts into the correct zone. Figure 6.16 shows the arrangement. As can be seen if, say, feeder 1 is connected to busbar section A then auxilliary contacts on that isolator connect the current transformers on the zone A buswires. If the feeder was switched to the reserve busbar, then the current transformer would be switched to the reserve buswire.

Although not shown in the diagram the isolator auxilliary contacts are usually duplicated, by connecting two in parallel, for security. In addition, normally closed isolator auxiliary contacts are used to short-circuit the current transformer when both isolators are open. This is to prevent damage to the CT should a fault develop between the CT and the isolators with the feeder energised from the remote end. Because the CT circuits are switched the integrity of the system in impaired and buswire supervision is considered essential.

MESH-CONNECTED SUBSTATION

A mesh-connected substation is shown in single-line form in Fig. 6.17. The protection consists of a fully discriminative scheme with high-impedance relays at each corner. A fault on any corner trips the two circuit-breakers associated with that corner and initiates an intertripping circuit to open the circuit-breaker(s) at the remote end(s).

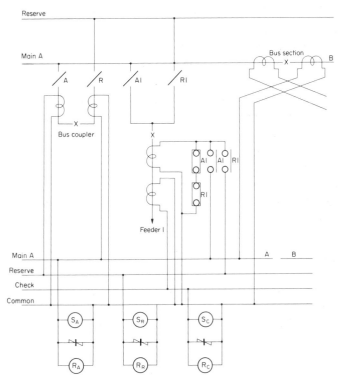

FIG. 6.16   PART OF A BUSBAR PROTECTION SCHEME FOR TWO MAIN BUSBARS
AND A RESERVE BUSBAR

As the clearance of a fault disconnects only a small amount of equipment the provision of a check scheme is unnecessary.

BIASED DIFFERENTIAL PROTECTION

Another variation of the Merz–Price principle is a scheme which counteracts the effect of spill current in the relay during through-faults by using the current circulating around the CT loop to restrain the relay. Figure 6.18(a) shows the circuit. R is the relay.

Under load or through-fault conditions current flows in the biasing coils to restrain the relay from operation and although a fairly large spill current may flow in the operating coil of the relay it will be well below the value required to overcome the restraining force. The strength of the scheme lies in the fact that when the spill current is likely to be high, the restraining force is at its greatest.

The most onerous condition under which the relay would be required to operate is an internal fault fed from one end only (Fig.

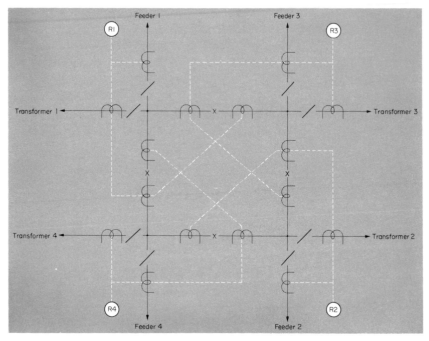

FIG. 6.17   MESH SUBSTATION BUSBAR PROTECTION

6.18(b)). Under this circumstance the operating current is the same as the restraining current but the operating force greatly exceeds the restraining force. When the system is fed from both ends (Fig. 6.18(c)) the net effect of the restraining force is reduced and the operating current is increased.

The ratio of operate/restraining current is usually expressed as a percentage. Figure 6.19 shows a relay with 20% bias or alternatively a bias slope of 20%. The line represents the boundary line of relay operation. Above the line the relay operates, below the line the relay restrains from operation. The vertical scale is the amount of current required in the operating coil (I_R) whilst the horizontal scale is the amount of current in the restraining coil (I_B). Or, referring to Fig. 6.18(a), $I_R = I_1 - I_2$ and $I_B = \frac{1}{2}(I_1 + I_2)$. There is a deviation from the true bias line particularly at the low end of the scale because the relay needs some current to overcome the control spring force. In Fig. 6.19 the setting is 40%.

The two types of relay used for this application are the induction disc relay and the axial moving-coil relay. The latter is used where high-speed operation, 60 ms, is required whilst the induction disc relay is used where a slower operating speed can be tolerated or where a time delay is required.

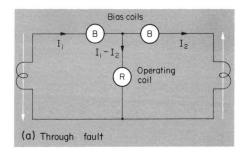

(a) Through fault

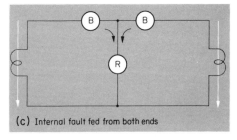

(b) Internal fault fed from one end only

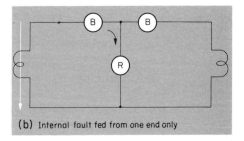

(c) Internal fault fed from both ends

FIG. 6.18   BIASED DIFFERENTIAL PROTECTION

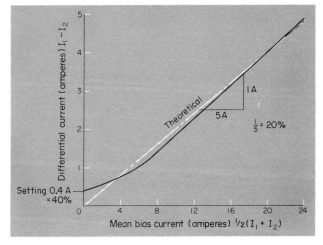

FIG. 6.19   BIASED DIFFERENTIAL RELAY: SETTING 40%, BIAS SLOPE 20%

The induction disc relay has two separate elements operating on a common disc. One element is arranged to move the disc in the operate direction whilst the other element, which has a centre-tapped winding, is arranged to move the disc in the restraining direction.

When no time delay is required the contact gap is small. In relays where time delay is available the contact gap is capable of being increased so that disc travel is increased.

Where axial moving-coil relays are used the relay has two coils each fed by a rectifier and a transformer. The moving-coil relay has a much lower setting than the induction-type relay and consequently the bias slope is much closer to the theoretical curve.

This type of protection is mainly used for the protection of transformers where there is difficulty in providing current transformers of the correct ratio because of tap changing. The following chapter considers the problem more closely.

Chapter 7

Transformer Protection

Possibly the most important item of equipment in an industrial power system is the transformer. These range in size from large incoming units which deliver power at the distribution voltage to those for low voltage utilisation and those for lighting systems.

As in all protection schemes the cost has to be related to the cost of the equipment it is protecting and transformers are no exception. It must be said, however, that in specifying a scheme the economic effect of the loss of the unit and the cost to repair a major breakdown should be taken into account. There is an arbitrary demarcation line at 33 kV. Where the HV winding is rated at a voltage above this high-speed protection can be justified. In all other cases the economic considerations mentioned above are pertinent.

TYPES OF FAULT

Because of its static nature the power transformer can be regarded as a very reliable unit. Nevertheless there is a possibility of failure because of internal faults as well as being subjected to stresses from external sources which could cause the internal fault condition.

Faults which are internal in origin are:

1. Failure of insulation of windings, laminations or core bolts—from damage in erection, inadequate quality or brittleness through ageing or overloading.
   Failure of the winding insulation resulting in inter-turn or earth faults. The possibility of a fault between phases is very small.
   Failure of the lamination or core-bolt insulation leading to increased eddy current causing heating of the core.
2. Oil deterioration which could be caused by poor-quality oil; penetration of moisture; decomposition because of overheating or the formation of sludge by oxidation as a result of bad electrical joints.

3. Loss of oil by leakage.
4. Inability to withstand fault stresses.
   This may be due to poor design or where repeated heavy currents set up severe mechanical stresses which causes packing and wedges to be loosened and finally shaken out.
5. Tap changer faults.
6. Cooling system faults.

External conditions which could cause faults to develop are:

1. Heavy through-faults.
   The high current would produce severe mechanical stress in the transformer windings and insulation.
2. Overloads.
   This would also produce mechanical stress in the windings and insulation and although these would be much less than under fault conditions they would be of much longer duration.
3. Switching surges.
   These surges, which may be several times the rated system voltage, have a very steep front and therefore a high equivalent frequency. This causes stress in the end turns of the winding and a risk of a partial winding flashover even though the insulation is usually reinforced in this area.
4. Lightning.
   This is only a risk where the transformer is connected to an overhead line and is usually protected by arrestors or spark gaps.

DIFFERENTIAL PROTECTION

Although unit protection is generally only applied to large transformers, which on most industrial installations would mean only the incoming supply transformers, it is considered first in the interest of continuity.

The basis of the schemes used for the overall differential protection of transformers is the Merz–Price system which relies for stability to through-faults on the balance of input and output current in the unit to be protected.

The input and output current of the transformer is, of course, substantially different but can be compensated for by having current transformers of different ratios on the primary and secondary sides. The CT ratios are arranged so that the secondary current of both sets of current transformers is always the same but because of the difference in primary current rating the magnetising characteristics of the two sets of current transformers would be quite different.

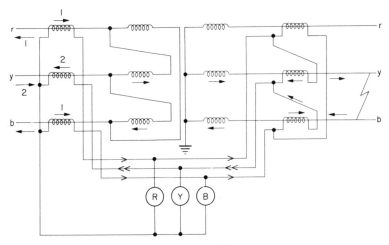

FIG. 7.1   DELTA/STAR TRANSFORMER. CURRENT DISTRIBUTION UNDER PHASE
FAULT CONDITIONS

In some cases there is a change in phase angle between primary and secondary currents, e.g. star/delta or delta/star transformers. This can be corrected in the CT secondary circuits by connecting the current transformers in the appropriate way. Where the transformer winding is connected in star the CT secondary windings should be connected in delta and vice versa. The secondary current rating of the current transformers to be connected in delta must be 0.577 times the rating selected for the relay and for the star-connected current transformers.

Figure 7.1 shows the connections for a delta/star transformer and the current distribution caused by a through phase fault. As can be seen the CT secondary current would satisfy the Merz–Price principle in that the output from each set of current transformers is the same in magnitude and phase. It should also be noted that the transformer current on the star side involves only Y- and B-phases and yet produces current in all three phases, on the transformer delta side the current in the Y-phase being double that in the R- and B-phases. This distribution, known as 2:1:1, has an important effect on some types of feeder protection.

MAGNETISING INRUSH

When an inductance is switched on to an a.c. supply when the voltage wave is at zero the so-called "doubling" effect occurs in which the first peak of the flux wave is greater than those of the steady-state condition. In the case of an iron-cored inductance, which is saturable and normally operates at the bend on the magnetising curve, the prospective doubling of the flux requires an enormous increase in the magnetising current. The effect is still further increased if the core contains initial remenance flux in the direction in which the first peak occurs. This phenomenon which occurs when transformers are switched to a supply is known as the magnetising inrush effect and initial currents as high at 14 times the rated current of the transformer are possible. The effect disappears in about 0.4 s.

As magnetising inrush current is associated with the saturation of iron the wave-form is very distorted and contains a large second harmonic. It flows in the primary winding of the transformer only and appears as operating current to a differential protection system. The operation is prevented under this condition by either a time delay or by using the second harmonic to restrain operation.

TAP-CHANGING

Most power transformers are provided with tappings so that the overall transformation ratio may be varied to suit the voltage requirements of the system.

As it is not practicable to vary the ratio of current transformers there will be out-of-balance, or spill, current in the differential relay when departure is made from the nominal transformation ratio. This means that plain circulating current scheme or a high-impedance relay is unsuitable for this application and a biased scheme must be used.

Unbiased schemes are sometimes used on small transformers—a circulating current scheme in conjunction with induction-type relays which have a time-delayed operation. To cater for the spill current, which flows at the extreme ends of the tap-changer range, settings higher than the transformer full-load current must be used. A modification of this system is known as the rough balance scheme. In this arrangement the current transformers are deliberately unbalanced by a small amount. Induction-type relays are used and the setting is chosen so that the same relay will provide overcurrent protection in addition to differential protection. For example, if an unbalance between current transformers of 20% is used and the relay has a setting of 25% then current equal to the setting will flow in the relay with a through-current of 125%. This is to say the response to overcurrents will be in all respects similar to that of a relay having a setting of 125%. The response to internal faults will be that of a differential system having a setting of 25%. The time setting of the relay must be adopted to co-ordinate with the system protection as a

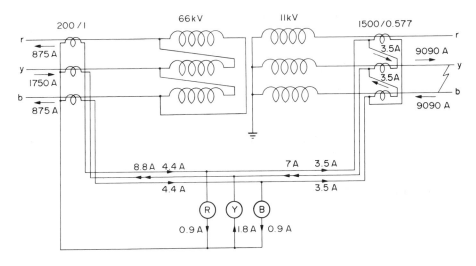

Fig. 7.2   Rough Balance Protection

whole and also deal with transformer magnetising inrush currents. Figure 7.2 shows the arrangement applied to a 20 MVA, 66/11-kV transformer.

The rated full-load current is 175 A on the 66-kV side and a CT ratio of 200/1 has been chosen. On the 11-kV side the full-load current is 1050 A and for balance the CT primary rating should be

$$\frac{66}{11} \times 200 = 1200 \text{ A}.$$

However, for the rough balance scheme 1500 A would be chosen and the secondary rating would be 0.577 A as they are delta connected.

The diagram shows the current flow to a fault on the system beyond the transformer. The relay plug setting would be 50% of 1 A and the time multiplier setting 0.21. This means that for the fault beyond the transformer the relay would receive a multiple of setting of

$$\frac{1.8 \text{ A}}{0.5 \text{ A}} = 3.6 \text{ times setting}$$

and the relay would operate in

$$\frac{3}{\log 3.6} \times 0.21 = 1.15 \text{ s},$$

whereas if the fault was inside the transformer protection zone the multiple of setting would be

$$\frac{8.8 \text{ A}}{0.5 \text{ A}} = 17.6 \times \text{setting}$$

and the relay would operate in

$$\frac{3}{\log 17.6} \times 0.21 = 0.5 \text{ s}.$$

Therefore the relay operates quickly for an internal fault but discriminates with other protection for a through-fault.

BIASED SYSTEMS

When the transformer is fitted with a tap changer it becomes necessary to employ bias. Bias also assists in countering CT inequalities but has no effect under inrush conditions.

Some schemes use induction relays in which both operating and restraining forces are produced by electromagnets, similar to induction overcurrent relays, wound to suit the particular function. The operation and restraining electromagnets operate on opposite sides of the same induction disc and both have tapped coils to provide a range of both setting and bias adjustment. The amount of disc travel can also be adjusted to provide a time-delayed operation to deal with the magnetising inrush condition. One of these relays would be required for each phase.

HIGH-SPEED BIASED SYSTEMS

The principle problem in providing high-speed transformer protection is the magnetising inrush current. To distinguish between this condition and other operating conditions use is made of the wave-form distortion which is characteristic of inrush currents.

Since the inrush phenomenon is due to saturation of the transformer iron, wave-form distortion is always present and the current contains a high proportion of harmonics. Analysis of the wave-forms has shown that the harmonic content expressed as a percentage of the fundamental varies from 20–30% at low values of inrush currents to the order of 60% at high values. The harmonic currents are filtered from the operating circuit and are passed through an additional restraining winding.

High-speed schemes generally use the permanent magnet moving-coil relay as the operating element. The moving coil has three separate windings for operation, restraint and harmonic restraint. As it is a d.c. device each winding is connected to a rectifier so that they

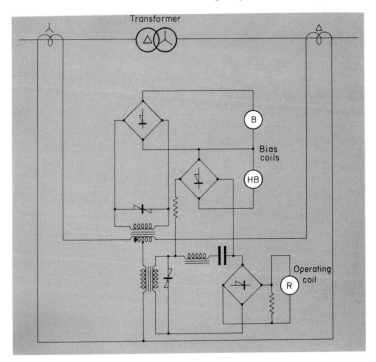

FIG. 7.3   SINGLE-LINE DIAGRAM OF HIGH-SPEED BIASED-DIFFERENTIAL
PROTECTION WITH HARMONIC BIAS

are supplied with currents of the correct polarity. Figure 7.3 shows the
connection. Two auxiliary current transformers are used to supply
the relay. One, with its primary winding connected to measure
circulating current, supplies the restraint circuit whilst the other
measures differential current and is connected to the operate circuit.
In both cases a non-linear resistor is connected across the auxiliary
current transformer secondary winding to protect the rectifier.
Adjustment of the bias section is by means of the variable resistor
connected across the main restraint coil. The range is from 0 to 40%
and will normally be set at a value 10% above the maximum percentage difference between the power transformer ratio at the
extremities of its tap range and the ratio of the current transformer
primary ratings.

A filter arrangement in series with the operating coil circuit filters
off the harmonic currents which are then passed through the harmonic
restraint coil. The harmonic circuit is proportioned so that 30% of
second harmonic will just balance the operating force produced by
the fundamental current. The amount of a higher harmonic which is
required to balance the fundamental is slightly less.

An instantaneous high-set overcurrent element is usually incorporated in the relay with a setting above the maximum magnetising inrush current.

EARTH-FAULT PROTECTION

Earth-fault protection using residually connected current transformers and high-impedance relays is frequently applied to one or both windings of a transformer.

The importance of this type of protection depends on the extent of protection provided by other protection. For example, if high-speed overall differential protection has been used then this will detect earth-faults and the fault will be cleared quickly. In this case the earth-fault relays are of secondary importance. If, however, the main protection is not high-speed then the earth-fault relays are essential if a fast clearance is to be achieved.

In many cases earth-fault current is low owing to the high impedance of the earth-fault circuit either by design—resistance earthed system, or because of earth-return conductor impedance— usually on low-voltage systems. Earth-fault level is also reduced if the fault is some distance down the winding from the terminals. The actual value depends on the winding connection and the method of earthing.

1. Star-connected Winding Earthed Through Resistance

An earth-fault on such a winding will result in a current which is related to the value of the earthing resistance chosen and is proportional to the distance of the fault from the neutral. The ratio of transformation between the primary winding and the short-circuited turns varies with the position of the fault which means the current into the transformer varies in proportion to the square of the proportion of winding short-circuited. This effect is illustrated (Fig. 7.4) from which it can be seen that with a basic setting of 15–20% for overall protection it is possible to protect only half the winding at most. In these circumstances restricted earth-fault protection must be considered essential.

2. Star Winding Solidly Earthed

In this case the actual fault current does not steadily decrease as the fault is moved towards the neutral point, but it is actually a maximum

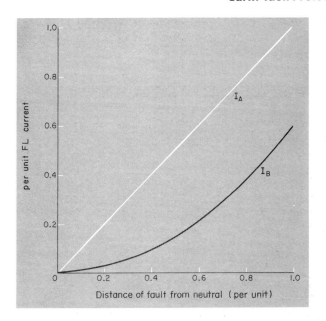

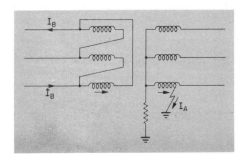

FIG. 7.4   FAULT CURRENT WITH A RESISTANCE EARTHED TRANSFORMER

at a point close to the neutral. The variations of fault and terminal current are illustrated in Fig. 7.5. Here the fault current seen in the line current transformers is consistently high enough for most overall schemes to be able to protect a satisfactory percentage of the winding. Nevertheless it is the practice to apply restricted earth-fault protection.

3. Delta Winding

No part of the delta winding operates at less than 50% of the phase voltage to earth and so the range of fault current occurring on such a winding is less than in case of the star winding. The actual value of

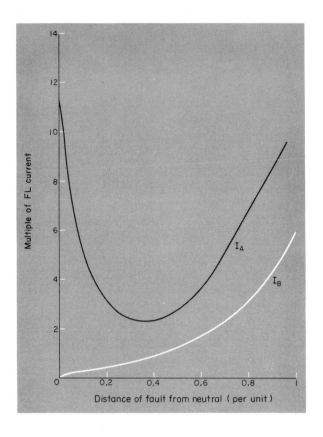

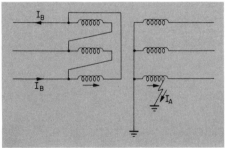

FIG. 7.5   FAULT CURRENT WITH A SOLIDLY EARTHED TRANSFORMER

fault current will, of course, still depend upon the manner in which the system is earthed and it should also be remembered that the impedance of a delta winding is particularly high to fault current flowing to a centrally placed fault on one leg. The impedance of the winding can be expected to be between 25 and 50% in this case.

PROTECTION OF AN EARTHED-STAR WINDING

The high-impedance relay is used with residually-connected current transformers, one in each phase and one in the neutral. The phase-connected current transformers detect any earth fault but a balancing current is supplied by the neutral current transformer when the fault is on part of the system other than the transformer, i.e. an external fault. This restricts the operation of the protection to faults within the transformer protection zone and it is known as restricted earth-fault protection.

Calculation of the setting voltage is as discussed in Chapter 6.

$$V_s = I_f(R_{CT} + R_L).$$

In general the largest value of $R_{CT} + R_L$ will be the neutral CT circuit. I_f is the maximum transformer three-phase through-fault current.

PROTECTION OF THE DELTA WINDING

Figure 7.6 shows a delta/star transformer with an earth-fault on the star side. The current on the delta side is equal and opposite in two phases and therefore the output of the residually connected current transformers will be zero as it will be under all load conditions. Only if there was an earth-fault on the delta winding would current flow into the circuit which was not balanced. The system is known as Balanced Earth-fault Protection.

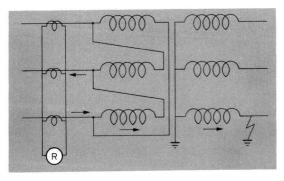

FIG. 7.6   DELTA/STAR TRANSFORMER. BALANCED EARTH-FAULT PROTECTION
FAULT ON STAR SIDE

Calculation of the setting voltage is as above but in this case I_f is the three-phase fault current at the primary terminals and R_L can be ignored.

COMBINATION OF OVERALL AND EARTH-FAULT SCHEMES

Where lead burdens are low it is possible to effect a saving of line current transformers by combining an earth-fault scheme with overall protection, in the manner indicated in Fig. 7.7.

A core balance-type auxiliary current transformer used with delta-connected main current transformers provides what is effectively a residual connection with regard to the restricted earth-fault protection.

Where the main current transformers are star connected an additional single element high-impedance relay is all that is required.

OPERATION LEVELS REQUIRED FOR RESTRICTED EARTH-FAULT SCHEMES

Where earth-fault current is limited by the inclusion of an earthing resistor, the effective operation level of the restricted earth-fault scheme is normally required to be less than 25% of the resistor rating.

Where the effective operation level is found to be less than, say, 10% of the rating of the transformer winding care should be taken to ensure that it is not possible for the charging currents from in-zone cables to cause instability.

STANDBY EARTH-FAULT PROTECTION

When a transformer is earthed through an earthing resistor or reactor it is usual to apply a time-delayed form of earth-fault protection having a characteristic to match the thermal rating of the earthing device. This relay which is generally regarded as the last line of defence and is intended to trip only in the event of a sustained earth-fault and will then trip both HV and LV sides of the transformer. In some cases two stages of earth-fault protection are applied and the LV and HV breakers are tripped in sequence.

BUCHHOLZ PROTECTION

Every type of fault which occurs under the oil in a transformer gives rise to generation of gas which may be slow in the case of minor or incipient faults and violent in the case of heavy faults. This fact is

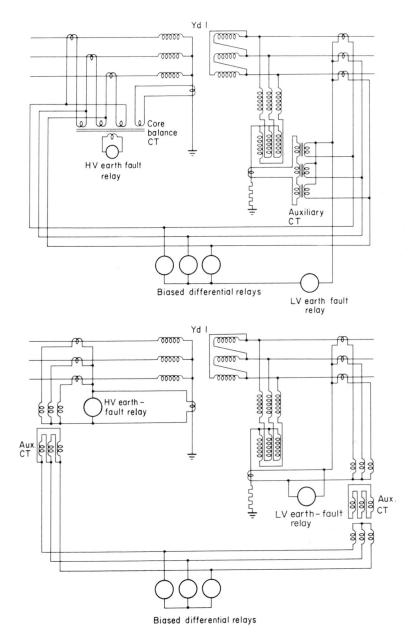

Fig. 7.7 BIASED DIFFERENTIAL AND EARTH-FAULT PROTECTION
OF POWER TRANSFORMERS

made use of in the Buchholz relay which is inserted in the pipe connection between the transformer tank and the conservator.

The Buchholz relay generally has two elements. A float switch which detects a fall in oil level and a combined hinged flap and float switch mounted so that it will detect any rapid movement of oil from the transformer to the conservator.

A slow generation of gas due to a minor fault causes a stream of bubbles which pass upwards towards the conservator, but are trapped in the Buchholz chamber, resulting in a slow fall of the oil level and consequent lowering of the upper float. This float is provided with a contact which would normally be connected to an auxiliary relay designated "Buchholz Gas" or "Buchholz Alarm".

When this occurs the gas sample should be collected by connection of a bottle with a flexible tube to the release cock. Information regarding the state of the transformer can be obtained from analysis of the gas and the rate at which it is generated.

In the main the relay will operate for one of three reasons:

1. Air inadvertantly introduced during filling or because of mechanical failure of the oil system.
2. Gas produced by the breakdown of the oil.
3. Gas produced by the breakdown of solid insulation.

The gases produced are hydrogen, acetylene and carbon monoxide and it is the proportions of these which point to the type of fault which is producing gas. Of these acetylene is soluble in oil and the presence or absence of this in the Buchholz relay chamber is not a good guide. Therefore the results depend on the gas-analysis equipment being capable of identifying hydrogen and carbon monoxide in quantities greater than 1%. In addition to the three gases mentioned small proportions of other gases, benzene, carbon dioxide and methane as well as nitrogen and oxygen, are also produced.

The information which can be derived from the analysis is as follows:

1. If the gas is mainly hydrogen with less than 2% carbon monoxide then the fault is likely to involve only the insulating oil.
2. If the gas is hydrogen with about 20% carbon monoxide then the fault is concerned with both solid insulation and insulating oil.

In the first case the transformer can be left in service providing that the accumulation of gas is slow. If successive alarms occur within a week then it should be taken out of service and examined.

In the second case where the burning of solid insulation is involved the transformer should be taken out of service irrespective of the time taken for the accumulation of gas.

A heavy fault would give rise to an explosive generation of gas which displaces the oil and causes a surge to pass along the pipe towards the conservator and in so doing displaces the flap switch which operates an auxiliary relay designated "Buchholz Surge" or "Buchholz Trip". Under these circumstances no attempt should be made to re-energise the transformer until an examination of the windings has been made.

A leakage of oil from the transformer tank causes a gradual fall in oil level which will in the first place operate the upper float to give an alarm and if not corrected will cause the lower float to fall and trip the circuit-breaker.

Special precautions are required when the transformer oil is being circulated for cleaning and also after the oil has been changed in the transformer because of trapped air bubbles which can displace the floats. At such times the device is usually connected for alarm only. The Buchholz relay is the best device available for detecting incipient faults and is specially sensitive to interturn faults.

OVERCURRENT PROTECTION

Distribution transformers are not usually fitted with overall differential protection and so it is necessary to provide overcurrent protection. The relays are generally of the induction type with the usual inverse definite–minimum time characteristic. The advantage of using this type of relay is simplicity but it has relatively high settings and long clearance times. If the possibility of phase faults within the transformer are considered to be remote then this may be acceptable, otherwise the risk of damage due to slow clearance times and the restricted amount of winding protected constitute serious limitations.

Overcurrent protection does, however, provide suitable back-up protection and here its characteristics are no disadvantage and as such it protects the system as a whole. In many installations it is used as primary protection for the LV busbar.

INSTANTANEOUS HIGH-SET OVERCURRENT PROTECTION

Mounted in the same case as the IDMT overcurrent relay is an instantaneous relay which takes advantage of the change in fault level between LV and HV sides of the transformer.

The impedance of a transformer is determined during the short-circuit test. In the transformer manufacturer's works two tests are conducted on all transformers. The open-circuit test in which nominal voltage is applied to the transformer on no-load. The current, this is the magnetising current, and the power input are measured. The power input is virtually the iron losses as current, and therefore the copper losses, are low.

The second test is the short-circuit test where a three-phase short-circuit is connected across the secondary winding and the voltage raised across the primary winding until full-load current flows. The voltage, which is the impedance voltage, and the power input are measured. In this case the losses are virtually the copper losses as the voltage, and therefore the iron losses, are low. The impedance voltage is usually expressed as a percentage of the nominal voltage. A typical value for a distribution transformer is 6% which means that with a short-circuited secondary winding an 11 kV transformer would require

$$\frac{6}{100} \times 11000 = 660 \text{ V}$$

across the primary winding to produce full-load current.

If the nominal voltage was applied to the primary winding with the secondary winding short-circuited the current which would flow in the transformer would be 100/6 = 16.7 times full-load current. Therefore any fault on the secondary side of the transformer cannot be greater than this and so if the high-set relay was set to operate at a value higher than 16.7 times full-load current it will not operate for a fault beyond the transformer and will therefore discriminate with all protection on the system beyond the transformer. The fault level on the HV side of the transformer will be far greater than the setting current and so this fault will be cleared quickly.

The instantaneous operation of the relay is also important in time-graded overcurrent schemes. If there was a ring system on the LV side of the transformer with relay settings producing relay operating times of 0.5, 0.9 and 1.3 s, then the transformer IDMT relay would be set to operate in 1.7 s, and any IDMT relay between the transformer and the source of supply would have to be set to operate in 2.1 s. With a high-set relay on the transformer the setting of this relay can be reduced to give an operating time of 0.5 s.

OVERLOAD PROTECTION

Large transformers are usually fitted with oil and winding temperature indicators. These each comprise a bulb of volatile liquid mounted in the hot oil near the top of the transformer tank and connected by capillary tubing to a Bourdon-type pressure indicator which is calibrated in temperature and has alarm contacts.

The winding-temperature indicator uses the principle of a thermal image. The bulb, in addition to being heated by hot oil, is also heated by a small heater energised by a current transformer which measures transformer current.

PROTECTION OF A TYPICAL INDUSTRIAL INSTALLATION

Two 4 MVA, 11/3.3 kV transformers which are part of a typical industrial installation are shown in Fig. 7.8. The complete installation would probably have at least two 11/0.415 kV transformers and a

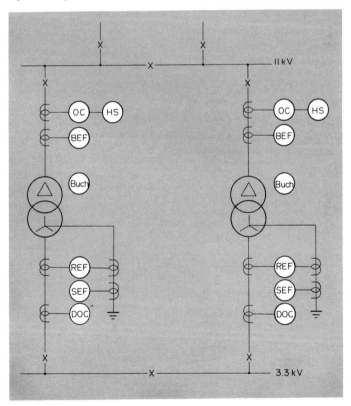

FIG. 7.8   PROTECTION OF TRANSFORMERS IN A TYPICAL SUBSTATION

B	Buchholz
BEF	Balanced Earth Fault
DOC	Directional Overcurrent
HS	Instantaneous Overcurrent
OC	IDMT Overcurrent
REF	Restricted Earth Fault
SEF	Standby Earth Fault

lighting transformer but, as the principles governing the application of the protection are similar, only the protection of that part of the installation shown is covered.

The 11 kV busbar fault level is 500 MVA and the transformer impedance is 6.4% on rating.

1. Fault at 3.3 kV busbar with two transformers, 1 MVA base.

Source impedance $= \dfrac{1}{500} = 0.002$ p.u.

Transformer impedance, $\dfrac{6.4}{100 \times 4} = 0.016$ p.u.

Total impedance, $0.002 + \dfrac{0.016}{2} = 0.01$ p.u.

Fault level $= \dfrac{1}{0.01} = 100$ MVA.

Each transformer 50 MVA

on 11 kV side, $I = \dfrac{50}{\sqrt{3} \times 11} = 2.62$ kA,

on 3.3 kV side, $I = 2.62 \times \dfrac{11}{3.3} = 8.75$ kA.

2. Fault at 3.3 kV busbar with one transformer.
   Total impedance $= 0.002 + 0.016 = 0.018$ p.u.
   Fault level $= 55.6$ MVA

on 11 kV side, $I = \dfrac{55.6}{\sqrt{3} \times 11} = 2.92$ kA,

on 3.3 kV side, $I = 2.92 \times \dfrac{11}{3.3} = 9.73$ kA.

Transformer FL current

on 11 kV side, $I = \dfrac{4000}{\sqrt{3} \times 11} = 210$ A,

on 3.3 kV side, $I = 210 \times \dfrac{11}{3.3} = 700$ A.

CT ratios
11 kV side 250/1   Resist 0.4 Ω   Balanced EF.
        250/1   Overcurrent.

3.3 kV side 800/1   Resist 2.23 Ω Restricted EF.
        800/1   Overcurrent.
3.3 kV neutral CT 800/1   Resist 2.0 Ω Restricted EF.
                800/1   Standby EF.
Resist of leads from neutral CT to relay 0.5 Ω.

RESTRICTED EARTH-FAULT PROTECTION

$$R_{\mathrm{CT}} + R_L = 2.0 + 0.5 = 2.5\ \Omega \text{ at } 20^{\circ}\mathrm{C}$$
$$= 2.5 \times 1.2 = 3\ \Omega \text{ at } 70^{\circ}\mathrm{C}.$$

$$I_f = \frac{9.73 \times 1000}{800} = 12.16\ \mathrm{A},$$

$$V_s = 12.16 \times 3 = 36.5\ \mathrm{V}$$

as the relay has a filter circuit, set to 40 V.

The magnetising curve (Fig. 7.9) shows the knee-point voltage in excess of 80 V and so the setting is satisfactory. If the knee-point

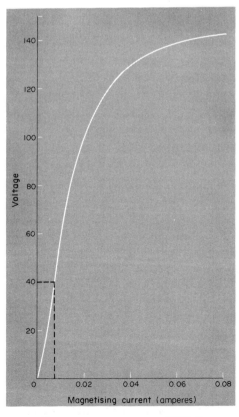

Fig. 7.9   Magnetising Characteristic of an 800/1 Current Transformer

voltage had been less than 80 V, say 66 V, then a setting of 30 V would have been chosen to ensure that the relay received at least twice its setting under fault conditions. This means, of course, that there is a risk of mal-operation during heavy through-faults but as stability can in general be guaranteed up to twice the fault current used in the calculation the risk is very small.

Current Setting

At 40 V the CT magnetising current is 0.008 A, the relay setting current is 0.02 A and the non-linear resistor current at 40 V is 0.001 A.
The overall setting is therefore

0.008 × 4 + 0.02 + 0.001 = 0.053 A

or in terms of primary current

800 × 0.053 = 42.4 A.

It may be considered that a setting of 0.053 A is too sensitive and a setting of 10% is preferred. In this case a resistor would have to be connected across the relay to shunt

$0.1 - 0.053 = 0.047$ A at 40 V, i.e. a resistor of $\dfrac{40}{0.047} = 850\,\Omega$.

BALANCED EARTH-FAULT PROTECTION

$R_{CT} = 0.4\,\Omega$ at 20°C
$= 0.4 \times 1.2 = 0.48\,\Omega$ at 70°.

$\text{Fault current} = \dfrac{500}{\sqrt{3} \times 11} = 26.2\,\text{kA}.$

$I_f = \dfrac{26.2 \times 1000}{250} = 105\,\text{A}.$

$V_S = 105 \times 0.48 = 50.4\,\text{V}.$
Set to 50 V.

The magnetising curve (Fig. 7.10) shows that the knee-point voltage is greater than 100 V and so the setting is satisfactory.

Current Setting

The current setting is the relay current, 0.02 A, plus the total CT magnetising current at 50 V, 3 × 0.023 A, and the current taken by the non-linear resistor, 0.001 A.

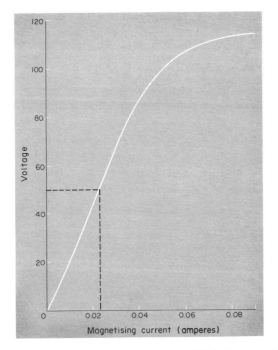

FIG. 7.10   MAGNETISING CHARACTERISTIC OF A 250/1 CURRENT TRANSFORMER

Current setting $= 0.02 + 3 \times 0.023 + 0.001 = 0.09$ A or in primary terms $250 \times 0.09 = 22.5$ A.

OVERCURRENT PROTECTION

The maximum through-fault current is 2920 A.
The instantaneous high-set relay must be set to about

$$1\tfrac{1}{3} \times \frac{2920}{250} = 15.6, \text{ set to } 16.$$

A fault on the 3.3 kV busbar would cause both IDMT overcurrent relays to operate to clear the fault. In the absence of any 3.3 kV busbar protection this has to be accepted. However, a phase fault between the transformer LV circuit-breaker and the transformer would also cause both transformers to be tripped. To prevent this happening directional overcurrent relays are used on the secondary side and will only operate when fault current flows into the transformer. These are set to operate faster than the 11 kV side overcurrent relays and therefore only the faulty transformer will be tripped.

The current setting of the directional relay should take into account the normal load current because, although this will not be in the direction of operation, if a setting lower than this is chosen the relay is overloaded and could in extreme cases burn out.

$$\frac{700}{800} \times 100 = 87.5\%, \text{ set to } 100\%.$$

The maximum fault current which can flow in this relay is 8750 A,

$$\frac{8750}{800} = 10.9.$$

Time for full travel $\dfrac{3}{\log 10.9} = 2.89 \text{ s}.$

To discriminate with the restricted earth-fault protection to avoid confusion with relay flags a discrimination time of 0.5 s should be used.

Time multiplier setting $= \dfrac{0.5}{2.89} = 0.173.$

Set relay to 100%, 0.18.

The relay on the 11 kV side is to discriminate with the directional overcurrent relay but also must discriminate with the 3.3 kV fuses. The relay must not operate when motors are started on the 3.3 kV side when only one transformer is in service.

Say the largest fuse on the 3.3 kV system is 160 A. The characteristic is shown in Fig. 7.11. The transformer full load current is 210 A so a setting greater than

$$\frac{210}{0.9 \times 250} \times 100 = 93.3\%, \text{ i.e. } 100\%, \text{ is required.}$$

The current setting is 250 A which when referred to the secondary side is 833 A. The characteristic curve of the relay plotted on the same graph as the fuse characteristic shows that the relay will discriminate with the fuse at all time multiplier settings.

The relay must discriminate with the directional relay at the maximum current which affects both relays. 8750 A at 3.3 kV which is 2620 A at 11 kV.

Multiple of setting $\dfrac{2650}{250} = 10.5.$

Time for full travel $= \dfrac{3}{\log 10.5} = 2.94 \text{ s}.$

Operating time of relay $= 0.5 + 0.4 = 0.9 \text{ s}.$

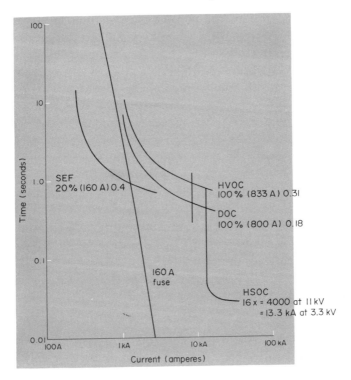

FIG. 7.11   TIME/CURRENT GRADING CURVES FOR 4 MVA TRANSFORMER

Time multiplier setting $\dfrac{0.9}{2.94} = 0.306$.

Set to 100%, 0.31.

STANDBY EARTH-FAULT PROTECTION

This relay was originally used to protect transformer neutral earthing resistors and to set it to match the thermal characteristics of these was straightforward. The major use now is as the last line of defence and as such it is required to detect and disconnect any low-level earth faults. It is difficult to determine a setting but it must be said that in the installation being considered it is not very important as all outgoing circuits are likely to be fitted with earth-fault relays. So in this case a setting of 20%, 0.4 will clear an earth-fault where the fault current is equal to transformer full-load current in about 10 s.

If an 11/0.415 kV system was being considered then the setting of relay becomes more important as some of the outgoing circuits may be protected only by fuses and have no earth-fault protection whatsoever. On a 415 V circuit many of the small motor drives will have only fuses and thermal overload relays. The standby earth-fault relay is required as far as possible to detect earth faults whose magnitude is insufficient to operate the motor protection. On the other hand, when the relay operates it results in complete loss of supply to the whole busbar and so unnecessary operation must be avoided.

A sensible approach to the problem would be to arrange the relay to discriminate with the largest fuses on the circuit which does not have earth-fault protection. A long-time relay is required.

Typically on a 415 V system all motors of 37 kW and greater will have earth-fault protection. The largest motor which has not is, say, 30 kW and say there is a 100 A fuse protecting this motor. Using a template of the stand-by earth-fault relay characteristic and moving it along the 125 A fuse curve to give a discrimination margin. The position which gives discrimination and a reasonable operating time at the transformer full-load current should be chosen, say 320 A, 0.1.

To summarise the settings would be:

11 kV side		
IDMT overcurrent	100%	0.31
Inst. HS overcurrent	16	
Balanced earth fault	40 V	
3.3 kV side		
Directional overcurrent	100%	0.18
Restricted earth fault	50 V	
Standby earth fault	20%	0.4

It is usual to arrange for all the above protection including the Buchholz surge relay to trip both the 11 kV and 3.3 kV circuit-breakers. A possible exception is the standby earth-fault protection which is sometimes used to trip only the 3.3 kV circuit-breaker or to trip the 3.3 kV side and then after a time delay, if the fault still persists, the 11 kV side.

As each relay has two contacts there are no problems associated with tripping both 11 kV and 3.3 kV circuit-breakers when the distance between them is small. However, where the 11 kV and 3.3 kV switchgear are at different locations an intertripping system must be used—see transformer feeders.

Chapter 8

Feeder Protection

DIFFERENTIAL PROTECTION

Any feeder protection scheme must have adequate speed of operation under fault conditions to minimise danger and discrimination in operation to localise the loss of supply.

In many circumstances these requirements are best met, or indeed only met, by the use of a unit scheme of differential protection for a particular section of feeder. For example, in a distribution system mainly protected by time-graded overcurrent protection, the minimum permissible operation time for a particular section may be unacceptably long.

Differential protection is principally of the longitudinal form, in which measuring equipment situated at each end of a feeder section is interconnected by means of pilot wires to compare the magnitude and phase angle of the current entering and leaving the feeder. Transverse differential schemes, in which currents in two or more parallel feeders are compared at one line end only, are suitable only where certain restrictive system conditions are met and are therefore rarely used.

Many schemes of differential protection for feeders are based on the Merz–Price system. In practice pilot connections in the role of CT leads would present an excessive burden and so a further stage of current transformation is introduced, also it is desirable for the purpose of tripping to have a relay at each line end. This means that the relay is not connected at an equipotential point and so to prevent operation because of this, and the inequalities of current transformer performance under heavy external fault conditions, a bias feature is generally included so that effective settings are increased as primary current magnitude is increased.

Thus most of the protective systems in use at present are derived from one of the basic arrangements shown in Fig. 8.1, the circulating current scheme or the balanced voltage scheme.

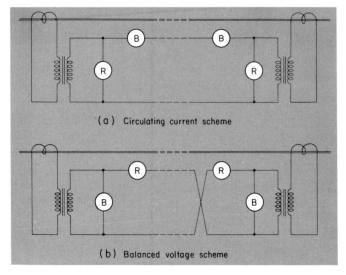

(a) Circulating current scheme

(b) Balanced voltage scheme

FIG. 8.1 LONGITUDINAL DIFFERENTIAL PROTECTION

FACTORS AFFECTING THE DESIGN

1. Type of Relay Element

Induction elements with a rotating disc have been widely used for schemes of medium operating speed. The bias feature is generally provided in such elements by the addition of a low-resistance "shading" loop to one arm of the electromagnet.

In high-speed systems the relay element is typically of the rotary or axial moving-coil pattern in which a coil winding, with two sections for operation and bias respectively, is operated between the poles of a permanent magnet. The relay is d.c. operated and the coils are fed from rectifiers.

2. Current Input Equipment

From an economic point of view it is desirable that any differential system of feeder protection should require only two pilot conductors if possible. This can be achieved only if the current input equipment for the relay elements has a means of reducing multiple input quantities from a three-phase primary system to a single signal for comparison in the pilot circuit.

The simplest method of producing a single-phase output is to use a summation primary winding on an interposing auxiliary current transformer. The sections of the primary winding have turns in the

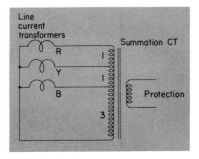

FIG. 8.2   SUMMATION CT CONNECTIONS

ratio of $1:1:n$ for R–Y, Y–B and B–N connections respectively where n may vary between 2 and 4 in general.

The summation transformer is a simple solution but its use results in the scheme having a different sensitivity to each type of fault. For example, the summation CT winding shown in Fig. 8.2 has primary winding turns in the ratio $1:1:3$. If the sensitivity of the scheme is such that, for R–Y phase fault, 50% of nominal current is required for operation then for an R–B phase fault 25% of nominal current would be required. That is because between R and B there are twice as many turns as between R and Y. Therefore only half the current is required.

The complete list of settings would be:

	Turns/R–Y turns	Setting
R–Y, Y–B	1	50%
R–B	2	25%
R–N	5	10%
Y–N	4	12.5%
B–N	3	16.7%

The three-phase setting will involve R–B turns but there will be a phase difference between R–Y and B–Y of 60°. This makes the effective turns/R–Y turns ratio 1.73 and therefore the three-phase setting is 50/1.73 = 28.9%.

For voltage-balance systems the auxiliary summation transformer may be of the quadrature pattern, which has an air gap in the core, such that a nominally linear relationship is provided between current input and voltage output. Alternatively, a conventional current transformer may be used in conjunction with an output resistor to provide a voltage reference at the relay.

With most protective systems it is necessary to limit the output voltage which can appear at the pilot terminations. This limitation may be controlled by the magnetisation characteristic of the summation current transformer and the addition of non-linear resistance across the transformer output terminals. It follows that at current levels in excess of that at which the limitation is imposed the current comparison to be achieved by the protective system will relate principally to phase angle and not current magnitude.

3. Pilot Characteristics

In most industrial distribution systems the pilot length will be very short and the effects of pilot resistance and capacitance and induced voltages in the pilots can be ignored completely. However, as many schemes include equipment to compensate the effects of long pilots it is necessary to examine these effects.

The wide variation of pilot wire characteristics presents a paramount problem in designing practical differential protection. In particular the presence of shunt capacitance introduces phase and magnitude differences in the two pilot currents. There are two main categories of pilot conductor, which may be distinguished by their resistance per unit length and their resistance/capacitance ratio.

(I) PILOTS HAVING A LOW RESISTANCE/CAPACITANCE RATIO

Generally cable having 2.5 mm^2 copper conductors is used for pilots in this category. The intercore capacitance per unit length of such pilots is relatively high and is generally the principle considera- tion in determining the maximum feeder length which can be pro- tected by schemes designed for such pilots. Practically all schemes for industrial distribution systems will be in this category.

(II) PILOTS HAVING A HIGH RESISTANCE/CAPACITANCE RATIO

These are in general telephone-type cables schemes using this type of pilot and will be limited to HV transmission lines.

In many protective systems the pilot resistance and capacitance is compensated to minimise errors.

(a) *Resistance.* With the simpler systems no attempt is made to offset variations in pilot resistance. Higher resistance merely increases the minimum operation level. In high-speed schemes a variable resistor is usually included in each relay so that the complete pilot circuit may be "padded" up to a prescribed nominal value thereby maintaining substantially constant operation levels for a wide range of pilot resistance.

(b) *Capacitance*. The principal effect of capacitance between pilot conductors is to provide a shunt path for current which would nominally be absent with ideally balanced voltage conditions on the pilots. The lowering of the shunt impedance by capacitance can be offset by the addition of a shunt reactor at each pilot terminal. When each reactor has been "tuned" to half the pilot capacitance at the fundamental frequency the high shunt impedance is substantially restored.

Compensation, however, is only partially achieved because the phase relationship between pilot terminal currents is not corrected by this method. Furthermore, the distributed nature of the pilot resistance and capacitance limits the degree of selectivity possible from tuning.

4. Insulation Requirements

When an earth-fault occurs on a feeder current flows down the faulty phase and an alternating magnetic field is produced around it. If the magnetic flux cuts any conductors, such as a pilot wire, a voltage is induced in it. The effect is worst with overhead lines because the return current, which would produce an opposing magnetic field, is some distance away flowing through the earth. In a cable system a large proportion of the return current would flow in the sheath and therefore the induced voltage is small. There is, of course, a difference in earth potential at the two locations which would also stress the insulation of the equipment connected to the pilot.

In a pair of pilot wires the same voltage is induced in each and so the voltage between the two pilot wires is practically zero—the induced voltage is between the two ends of the pilot. To prevent the induced voltage from circulating a current the pilot circuit is isolated from earth and all equipment connected to the pilot must be insulated from earth at a level to prevent damage from induced voltages.

The protection equipment is generally designed to comply with one or two insulation levels, 5 kV or 15 kV. Where the 15 kV level is to be met a suitable insulation barrier may be provided between the relays and pilots by interposing an isolating transformer at each pilot termination. Alternatively, the insulation barrier may be introduced between the primary and secondary windings of the summation current transformer, in this case the relays are insulated to withstand 15 kV to earth. Equipment for use on distribution systems requires only 5 kV insulation.

5. Sensitivity

The conventional connections to a summation transformer give the lowest settings for earth-faults, typically within the range of 10% to 40% of rated current. Phase-fault settings are higher, the relationship being dependent on the turns ratio of the primary winding of the summation transformer.

The use of bias in protective systems results in higher settings when through-load current is flowing to a level which depends on the summation CT ratio and the bias characteristic. The settings are raised by an amount represented by the bias but a setting that might be dangerously low with an unbiased scheme can be used.

6. Stability Requirements

(a) *Through-fault conditions.* Stability of the protection for through-fault conditions is assured by compliance with the basic principles of the system, circulating current or balanced voltage, assisted by the bias feature.

There is a problem under three-phase fault conditions because of "spill" current which flows in the CT neutral connection and therefore through the major portion of the primary winding of the summation CT. The spill current can be reduced and therefore the three-phase stability improved by the inclusion of a "stabilising" resistor in series with the neutral connection. The additional burden of this resistor tends to lower the stability level of the scheme for earth-fault conditions but a value is chosen so that similar stability levels are attained for earth and phase faults, typically 20 to 30× rated current.

With most schemes the stability level attainable falls in relation to length of the pilot circuit but this is generally acceptable because where a pilot circuit is long, the associated feeder is likely to be relatively long also and thus the possible through-fault current level is reduced.

(b) *Magnetising inrush condition.* Feeders, and in particular transformer-feeders, may be subjected to magnetising inrush current when power transformers are energised. The magnetising current will flow through each terminal of the feeder but may give rise to unbalanced currents from the main CTs of the feeder protection. For this reason a harmonic bias feature is added to some protective systems.

(c) *Line charging current.* The charging current drawn by a feeder may flow from one end only and is therefore capable of unbalancing a protection system and could lead to tripping. With overhead transmission lines the charging current is well below protection-operation levels and can therefore be neglected. Cables, however, have much higher charging current levels and this may determine the minimum permissible operation levels for the protection to ensure stability. A means must be provided in the relays for increasing the minimum setting when necessary.

PRACTICAL SYSTEMS

Figure 8.3 shows one end of a typical high-speed differential system. The current input is from three current transformers to the summation transformer with a stabilising resistor in the CT neutral connection.

A resistor is connected across the summation CT secondary winding which provides the reference voltage. During heavy through-faults the reference voltage is limited by the non-linear resistor so that the comparison of the input and output at the two feeder ends is by phase only rather than by magnitude and phase.

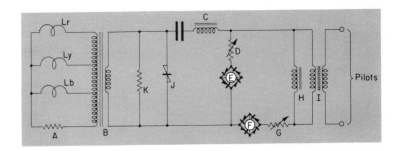

Fig. 8.3   Balanced Voltage Differential Protection

A	Stabilising resistor
B	Summation transformer
C	Tuned circuit
D	Bias adjustment
E	Relay bias coil
F	Relay operating coil
G	Pilot padding resistor
H	Pilot tuning choke
I	Isolating transformer
J	Non-linear resistor
K	Reference resistor
Lr, Ly, Lb	Line current transformers

PLATE 8.1 TRANSLAY RELAY (GEC Measurements)

The bias relationship is set by adjustment of the resistance in series with the bias coil. The tuned circuits (C) serve to block discharge of energy from the pilots and isolation transformers on clearance of a fault and diverting discharge current through both the bias and operation coils to maintain stability. The "padding" resistors (G) are adjusted so that the total circuit loop resistance between the relays at each line end is maintained at the designed value irrespective of different pilot resistances.

Feeders and interconnectors in distribution systems are usually protected by a simpler scheme having a medium speed of operation. The Translay system is a well-established example.

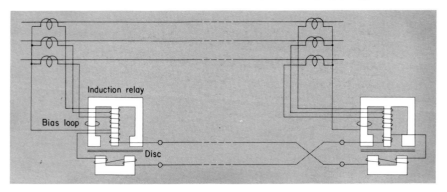

FIG. 8.4   TRANSLAY FEEDER PROTECTION

The scheme, which is shown in Fig. 8.4, uses two induction pattern relays. The upper electromagnet is used as a summation transformer with leakage flux to polarise the relay. Voltage produced across the secondary winding is nominally proportional to the input current and provides the voltage reference for the pilot circuit. The coils of the lower electromagnets of each relay are connected in series with the pilots. A shading ring, in the form of short-circuited turn, is fitted to one limb of the upper electromagnet to provide a bias feature. This type of bias has the advantage that when the through-fault current, and therefore "spill" current, is high the bias is also high.

The vector diagram of the system is shown in Figs. 8.5(a) and 8.5(b). The former showing the situation during a through-fault; the latter during an internal fault. In the through-fault condition the voltages induced in the secondary windings on the upper electro-magnets at each line end are opposed and nominally equal and the current in the pilots owing to any slight difference will be small. The voltage, however, will supply a leading current to the pilot wire capacitance but as the upper electromagnet flux is modified by the current flowing in the shading ring the capacitive current will restrain the relay as the lower coil flux leads the upper coil flux.

During an internal fault there will be a difference in voltage at each end and current will circulate and produce a flux in the lower electromagnet in such a direction as to cause operation.

Another balanced voltage scheme is the well-known Solkor R scheme which is shown in Fig. 8.6.

Resistors, each approximately equal to the pilot resistance and bridged by a rectifier, are connected in series with the pilot circuit to obtain an artificial mid-point connection for the relay coil.

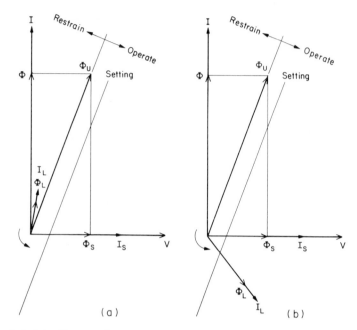

FIG. 8.5. (a) VECTOR DIAGRAM DURING THROUGH-FAULT CONDITIONS.
(b) VECTOR DIAGRAM DURING INTERNAL FAULT CONDITIONS

VECTOR DIAGRAMS FOR TRANSLAY PROTECTION

I   Upper electromagnet primary current

Φ Upper electromagnet flux produced by the primary current

V   Secondary voltage induced by Φ

I_L Lower electromagnet current

Φ_L Lower electromagnet flux

I_s Current induced in shading ring

Φ_s Flux produced by I_s

Φ_u Upper electromagnet output flux

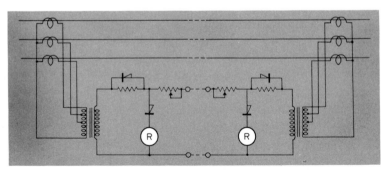

FIG. 8.6 SOLKOR R PROTECTION

PLATE 8.2   SOLKOR RELAY (Reyrolle Protection)

As the conducting directions of these rectifiers are in opposition, the equipotential point alternates between the two ends during through-fault conditions, each relay being at the mid-point on alternate half-cycles. The series rectifier in the relay coil circuit prevents current from flowing in the relay circuit when it is not operating at the electrical mid-point.

Under internal fault conditions current flows in the relay coils causing operation. Padding resistors are included to enable the resistance of the complete pilot circuit between the relays to be adjusted to 1000 Ω.

TEED FEEDER DIFFERENTIAL PROTECTION

The major difficulty which is encountered by any teed feeder protection scheme is the lack of bias at any end where the current flow to a through-fault is light compared to the flow at other ends. The heavy fault would result in "spill" which could cause mal-operation where the bias is low. In addition the output voltage from each end must be linear up to maximum fault level so that balance is always achieved irrespective of the distribution of the current at the three ends.

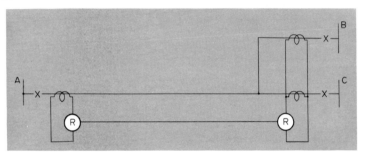

FIG. 8.7   TEED FEEDER DIFFERENTIAL PROTECTION

In some applications where two ends of a teed feeder are in the same location the two sets of current transformers can be paralleled and connected to a single relay of a plain feeder-protection scheme as shown in Fig. 8.7. This system is satisfactory providing there is no possibility of fault current flowing from B to C or C to B which would result in a high "spill" current with no bias.

TRANSFORMER-FEEDER PROTECTION

The transformer feeder is used in many industrial applications where a bulk power supply is required at a lower voltage at a remote point. The HV switchgear is installed at the local substation and the HV feeder is terminated in a transformer at the remote end where the LV switchgear is installed.

Overall differential protection can be used operating on the balanced voltage principle but the scheme is virtually two separate schemes because of the possible distribution of fault current in the summation CT during certain phase faults as illustrated in Fig. 8.8. As can be seen the phase fault shown will result in a $1:2:1$ current distribution in the current transformers. If this current was fed to the

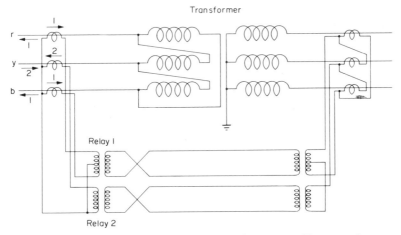

FIG. 8.8   TRANSFORMER FEEDER PROTECTION. BALANCED VOLTAGE SCHEME

summation CT there would be no output and therefore no bias which could result in relay operation because of spill current. The solution is to have two relays connected as shown.

INTERTRIPPING

Intertripping is required where it is necessary to operate a circuit-breaker which is remote from the relaying point.

The necessity arises because it is not possible to carry circuit-breaker trip coil current over a great distance by means of normal pilot wires. A circuit-breaker trip coil current is typically 10 A and therefore if tripping were attempted over a pilot of, say, 10 Ω resistance there would be a drastic reduction in the voltage at the trip coil.

The simplest intertripping scheme consists of a relay adjacent to the circuit-breaker and energised over pilot wires by the contacts of the protection relay. The relay-operating current would be very small and therefore the voltage drop in the pilots would be small.

Intertripping schemes are used principally with transformer feeders or on feeder differential protection schemes which do not have an inherent intertripping capability. An inherent intertripping capability is where, for any internal fault, both ends will trip even though there is no current at one end. Some schemes, notably Translay protection, require a current at each end to polarise the relay and is therefore not capable of tripping both ends unless current flows from both ends.

The protection pilots can, however, be also used for intertripping

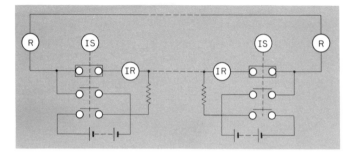

FIG. 8.9   INTERTRIPPING USING PROTECTION PILOTS

R   Protection relay secondary coil
IR   Intertrip receive relay coil
IS   Intertrip send relay coil

Normally closed contact

Normally open contact

in the manner shown in Fig. 8.9. Low-resistance intertrip receive relays are connected permanently in the protection pilot circuit without affecting the performance of the protection. Conversely the protection relay coils can withstand the passage of d.c. signals injected from the intertripping battery by means of an intertrip send relay.

Where transformer feeders are involved in some cases the feeder and the transformer are protected separately. Two sets of current transformers are mounted on the HV side of the transformer, one set for the feeder differential protection and one set for the transformer differential protection. If a fault occurs in the transformer then the LV circuit-breaker is tripped by the transformer protection but the fault will be seen as a through-fault by the feeder protection and therefore the HV circuit-breaker will not trip. A method which can be used for intertripping in this case is the short-circuiting of the feeder protection pilots to cause unbalance and therefore operation of the relay at the HV circuit-breaker.

Where there is no feeder differential protection it is still necessary to trip the HV circuit-breaker and a normal intertripping scheme is necessary. It is also necessary, of course, to trip the LV circuit-breaker when the protection at the HV circuit-breaker operates. This could be accomplished by two intertripping schemes involving three or four pilot wires. It is more usual to have a two-way intertripping scheme using two pilots as shown in Fig. 8.10. The arrangement requires switching of the pilots and therefore a multi-contact intertrip send relay is used in addition to the intertrip receive relay at each end of the pilot.

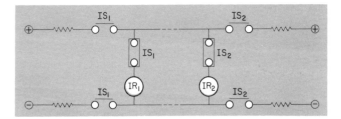

FIG. 8.10   TWO-WAY INTERTRIPPING

Although not relevant on distribution systems, where high voltages can be induced in the pilot wires special precautions need to be taken. As the intertrip receive relay is fairly sensitive it is necessary to surge-proof it against currents which could flow when a high voltage is induced. The surge-proofing takes the form of filter circuits which make the relay insensitive to circulating a.c. current and yet allows the relay to accept a low d.c. intertripping signal.

AUXILIARY EQUIPMENT

1. Check or Starting Relays

Where the three-phase setting of a differential protection system is less than load current, failure of the pilot circuit can lead to unnecessary operation of the protection. Thus where the integrity of the pilot circuit cannot be assured it is general practice to use starting or check relays in conjunction with the main protection so that the effective three-phase setting exceeds the maximum load level. The check or starting relays are usually of the attracted-armature pattern, with coils connected in the red, blue and neutral secondary connections to the main CTs.

Starting relays are relays which are arranged, when quiescent, to prevent the functioning of the main protection typically by short-circuiting the operation circuit of the latter by means of normally closed contacts. Check relays have their contacts in series with the tripping contacts of the main protection.

2. Pilot Supervision Equipment

Figure 8.11 shows a typical scheme. A low-level d.c. current is injected at end A and detected at end B. At end A a transformer rectifier unit is fed from the 240-V 50-Hz supply. The output is smoothed by the filter unit and the relay detects the loss of the supervision supply.

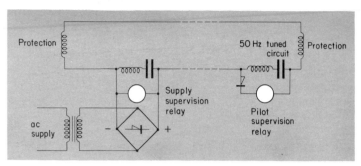

FIG. 8.11 PILOT WIRE SUPERVISION BY D.C. INJECTION

A similar alarm relay, suitably polarised, is connected in the pilot circuit at the other line end so that open-circuited, short-circuited or crossed pilot conditions may be detected. Simple time-delay elements are included at each line end to prevent alarms being given during system fault conditions. A disadvantage of this system is that the alarm relay is at a remote point. Supervision schemes are available where the pilot circuit is monitored by the Wheatstone bridge technique, with the pilot loop as one arm of the bridge. A detector operates if the pilot loop resistance changed by more than a predetermined margin which is adjustable between ±5% and ±20%.

In most industrial systems there is usually no shortage of pilot wires and it is sufficient to monitor that the pilot multicore cable is intact. This can be achieved by monitoring spare cores in the cable.

IMPEDANCE PROTECTION

Although impedance protection is not applicable to distribution systems to complete the picture the principles involved are worth consideration.

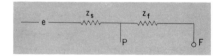

FIG. 8.12

An impedance diagram of a feeder fed from one end is shown in Fig. 8.12. Z_S is the impedance of the system between the source and the busbar to which the feeder is connected and Z_f is the impedance of the feeder. If a fault occurred in position F the current would be

$$I = \frac{e}{Z_S + Z_f}$$

and the voltage at the relaying point P would be

$$V = IZ_f = \frac{e}{Z_S + Z_f} Z_f.$$

The voltage and current would be measured at P by an impedance relay.

The basic operating principle of an impedance relay can be described by the operation of a balanced-beam relay—a type of relay which is no longer used. The relay is shown in Fig. 8.13. Two solenoids act on the beam and when the forces produced by them are equal the beam is balanced. The beam moves towards the solenoid producing the highest force.

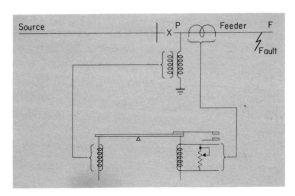

Fig. 8.13   Balanced-beam Relay

If one coil is connected to the voltage supply and the other connected in the current circuit the variable resistor can be adjusted so that balance is achieved with the voltage and current produced for a fault at F. If the fault is closer to P than F as the current increases, the voltage decreases and the beam moves towards the current operated solenoid. If the fault is beyond F the current decreases, the voltage increases and the beam moves towards the voltage solenoid. Therefore if the impedance from the relaying point to the fault is less than the set value the contact closes; if the impedance from the relaying point to the fault is greater than the set value the contact remains open.

The forces produced are dependent only on the magnitude of the voltage and current and are unaffected by the phase angle which

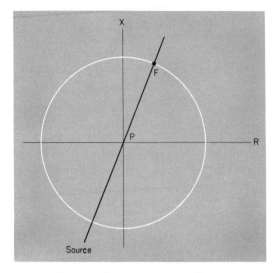

FIG. 8.14   OPERATING ZONE OF BALANCED-BEAM RELAY

means that the relay is non-directional and will operate for all values of impedance less than the setting impedance. The characteristic shown on an Argand diagram is a circle with the centre at the relaying point (Fig. 8.14).

This relay as well as operating for faults between *P* and *F* would also operate for any fault of less impedance than its setting in the opposite direction, assuming current would flow through the relaying point to the fault in this case.

This is clearly undesirable and in the early schemes this problem was overcome by the use of a directional relay which only allowed the impedance relay to operate if the fault was in the direction of the feeder which the relay was protecting.

Modern schemes use relays which combine the voltage and current in such a way to produce the Mho characteristic. On an Argand diagram this is a circular characteristic with the circumference passing through the origin and the diameter from the origin at an angle similar to the phase angle of the fault current (Fig. 8.15). The Mho characteristic is directional and the diameter can be adjusted to suit the impedance of the particular feeder with which it is associated.

In an ideal scheme the relay would be set to have a balance point at the end of the feeder which it is protecting. However, the relay cannot be set with such accuracy and there is a danger that it may "see" faults in the next feeder. This must be avoided and to ensure that it does not

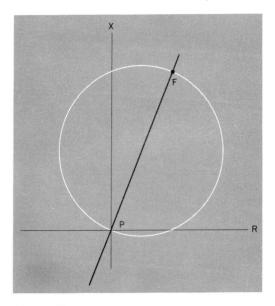

FIG. 8.15   MHO RELAY CHARACTERISTIC

happen the relay is set to have a balance point at about 15% short of the full feeder length. To protect this 15% and to discriminate with the protection on the next feeder a time delay is introduced after which the balance point of the relay is changed to include the rest of the feeder. It also includes some of the next feeder where it will act as a back-up protection. After a further time delay the balance point is again extended to cover the rest of the next feeder to provide complete back-up protection. Figure 8.16 shows how discrimination by time is achieved, and Fig. 8.17 shows the relay impedance charac-teristics. The successive changes in relay balance point are known as zones 1, 2 and 3.

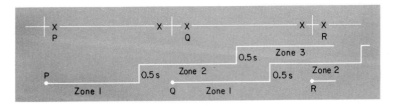

FIG. 8.16   IMPEDANCE PROTECTION-DISCRIMINATION BY TIME

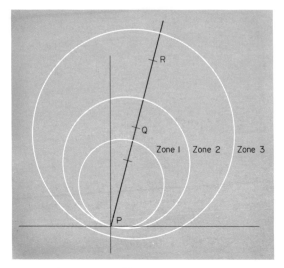

FIG. 8.17   MHO RELAY IMPEDANCE ZONES

It will be noted that the zone 3 characteristic differs from zone 1 and 2 in that the circumference does not pass through the origin. This is known as an offset Mho characteristic and is used to allow the first zone relay to measure and to start the timing sequence. The reason for the offset is so that a close-up fault is not on the edge of its characteristic as is the case with the normal Mho relay.

Impedance protection is applied to long feeders where pilot-wire protection is not feasible.

Chapter 9

Motor Protection

For many years a comprehensive system of motor protection has been considered essential for vital services and industrial processes to safeguard machines and their cables from damage caused by over-currents. Mechanical overload, stalling, single-phasing and short-circuits are some of the potential hazards. All of these result in overcurrents and are usually detected by either time delayed or instantaneous overcurrent relays.

The present day tendency is to employ motors to the limit of their thermal margins and to cater for this a relay with an inverse-time characteristic, similar to the thermal time characteristic of the motor, is used. The characteristic must allow the motor starting current to flow for a time in excess of the motor starting time.

It may be that a short resumé of the operation of the three-phase induction motor would be useful at this point. The three-phase voltage produces current in the stator winding which sets up a rotating magnetic field. This field flux cuts the short-circuited rotor conductors and induces a current in them. The interaction of the current and flux produces a torque which causes rotation. Figure 9.1 shows a torque-speed curve for a typical motor and superimposed on this curve is the torque-speed curve of a fan. Underneath is the current-speed curve for the motor. As can be seen the current is at starting current level until about 80% speed is reached.

The torque increases until it reaches a maximum, in this case at 90% speed, and the value at this point is known as pull-out torque. A further increase in speed causes the torque to decrease until it would become zero if 100% speed could be reached. At zero speed the torque is in excess of that demanded by the fan and therefore the motor accelerates. The speed increases steadily as the excess torque is roughly the same value up to 30% speed.

After 40% speed there is a large excess of torque and so the machine accelerates quickly until it delivers the amount of torque required by the fan at A. If the fan dampers were closed then the required torque is far less; the excess torque and therefore the

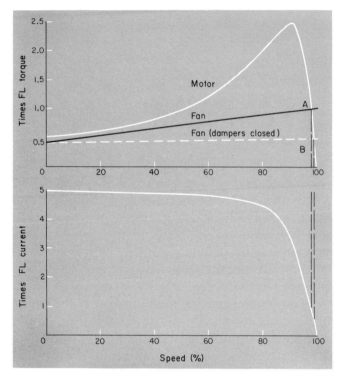

FIG. 9.1   TORQUE-SPEED AND CURRENT-SPEED CHARACTERISTICS OF AN
INDUCTION MOTOR

acceleration is greater and the machine runs up to speeds quicker and
delivers the amount of torque required by the fan at B. If vertical lines
are dropped from points A and B the load current will be indicated on
the current-speed curve. The actual current is greater than that
indicated which does not include the magnetising current.

From the torque-speed curve it can be seen that the risk of stalling
is greatest up to 30% speed where the difference between the motor
and load torque is least. If there was a reduction in motor torque,
which could happen if the voltage was depressed, to a level where it
equalled the fan torque at that speed then the motor would not
accelerate and would draw starting current.

The speed at which pull-out torque occurs depends on the ratio of
rotor resistance to rotor reactance. Rotor reactance is proportional to
the rotor frequency which in turn depends on the difference between
the speed of the rotor and the speed of the rotating field which has
been produced in the stator. This difference is the slip. Therefore
rotor reactance is proportional to slip frequency.

Pull-out torque occurs when Rotor Resistance = Rotor Reactance
when $R_2 = SX_2$, where S is the slip.

In the case shown in the curve the X/R ratio is 10/1 and therefore the pull-out torque will occur at

$$S = \frac{1}{10} = 0.1.$$

With slip-ring motors it is possible to introduce resistance into the rotor by connecting in a resistance bank. This will change the position of the pull-out torque. For example, if resistance is added so that the total resistance is equal to the reactance at 50 Hz then the pull-out torque will occur at $S = R/X = 1$, i.e. at motor standstill. This will produce a relatively high torque to accelerate the rotor quickly but with the load shown would run at only 82% speed.

If, when the motor achieved 30% speed, the value of external resistance was reduced to just below half then the pull-out torque would occur at $S = 0.5$, i.e. half speed. Finally the external resistor would be reduced to zero and the condition would be as shown in Fig. 9.1. The torque-speed curves for the three resistance steps are shown in Fig. 9.2.

The above is an accepted method of starting slip-ring motors but the change in external resistance values would be carried out smoothly to give the best acceleration.

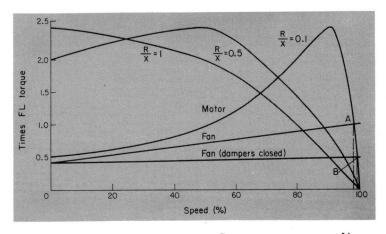

FIG. 9.2   INDUCTION MOTOR TORQUE-SPEED CHARACTERISTICS WITH VARIOUS R/X RATIOS

OVERLOAD PROTECTION

The most widely used relay for a.c. motor protection is the P & B Gold's relay which consists of three heaters supplied by three current transformers measuring stator current. The heaters are in the proxim-

PLATE 9.1   P & B GOLDS' MOTOR PROTECTION RELAY
(P & B Engineering Co. Ltd.)

ity of bimetallic strips which when heated produce a torque to move
the relay contacts towards a fixed contact. The deflection is proportional to current squared and therefore a motor operating at full load
would move the contact three-quarters of the way towards a final
contact set at 115%. Hence the operating time would only be a
quarter of the time required to operate the relay if the motor was
running light. Whilst this is obviously a desirable feature it should be
remembered that the thermal time constants of the bimetal and
motor differ widely and if the motor load is varying the bimetal will
respond in seconds, whereas the motor temperature change will take
minutes or even hours.

Because the bimetallic strips are heated indirectly, i.e. by heat radiated from the heaters as opposed to heating by passing current through the bimetal, there is a time delay in its response. This means that the moving contact will continue to move towards tripping even after the motor-starting current has disappeared. To avoid operation the relay operating time must be at least twice the motor starting time.

The relay has a special contact arrangement which operates if the current in any phase differs from the current in the other phases by more than 12%.

The reason that unbalanced phase currents require disconnection of the machine is that any unbalance in the current results in a negative phase sequence component which produces a rotating field in the opposite direction to the rotating field produced by the applied system voltage. This counter-rotating field will cause induced currents in the rotor of almost twice normal system frequency, resulting in overheating and possible damage.

Apart from the condition where one phase of the supply is missing completely, for example, owing to a blown fuse it may be thought that any unbalance in the system is small. This may be true in terms of voltage but as the negative phase sequence voltage will be applied to the standstill impedance of the motor the current will be substantial. If the negative phase sequence voltage was 5% then, if the starting current is 6 times full-load current, the negative phase sequence current will be 30%. In the case where one phase is missing completely the positive and negative sequence currents will be the same.

Other conditions which can cause unbalanced voltages are heavy single-phase loading or a blown fuse in a power factor correction capacitor circuit.

Overload and unbalanced load are conditions associated with the situation external to the motor. Overload is caused by an increase in the mechanical load whereas unbalanced currents are caused by the supply.

If tripping has been initiated by either of these conditions, indicated by operation of the flag relay marked "Thermal", it is usually in order to restart the motor. Only one restart should be attempted and during the starting period the relay should be used to diagnose the type of fault, overload or unbalanced current if it is still present.

As shown (Fig. 9.1) the starting current of a direct-on-line motor is practically constant at short-circuit level during most of the run-up period and there is, therefore, no means of detecting a stalled condition by current level alone. The thermal relay will trip the motor eventually, but because the time is long it may be too slow to prevent damage. In this case a single element stalling relay is used. This relay has a directly heated bimetal which gives a low overrun and therefore the operating time can be set close to the maximum run-up time. In some cases the run-up time is greater than the allowable stall time— this means that the condition can only be resolved by the addition of a speed-measuring relay.

INSULATION FAILURE

Protection against short-circuits is accomplished by instantaneous overcurrent relays but as these must have a setting greater than the starting current of the motor only a limited amount of the stator winding is protected. For example, if a short-circuit occurs at the motor terminals the current is the full system fault current whereas if the short-circuit was at the star point no fault current would flow as the star point itself is a short-circuit. Therefore along the length of the winding there is a reduction of fault current from maximum to zero. The decrease is not linear but roughly proportional to the square of the distance from the star point. Figure 9.3 shows this and also that an instantaneous relay which is set to avoid operation under starting conditions will only operate for faults just beyond the motor terminals. In fact these overcurrent relays can only be regarded as protection for the cable and the motor terminal box.

In a delta-connected motor there is, of course, no point in the winding where a short-circuit would produce zero current. Neverthe-less there is a large reduction in fault current for short-circuits away from the motor terminals.

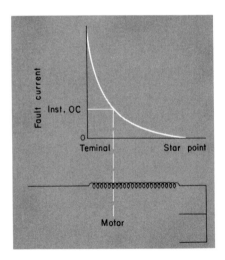

Fig. 9.3   Reduction of Fault Level in a Star Winding

The detection of short-circuits between windings is accomplished by the earth-fault relay. This is residually connected and has a setting of 10% which means that practically all of a star-connected winding will be protected. It is assumed that any insulation failure will result in an earth fault which is a reasonable assumption in that all the stator windings are in the close proximity to the stator iron.

The relay itself should have a voltage setting with a stabilising resistor but in some cases this is not required as stability is required only to the level of the motor starting current.

Operation of either the instantaneous overcurrent or the earth-fault relays operates a flag indicator marked "Instantaneous". When the motor has been tripped by these relays no attempt must be made to restart until an insulation test has been carried out.

SETTINGS

The P & B Gold's motor protection relay is available with different time ranges, namely 14, 20 and 30 minutes. These do not describe the relay in a very precise manner as they refer to points on the curves which are assymptotic. The times would have more meaning if they were quoted at a higher multiple of setting current. The 14- and 20-minute relays are the standard. A 30-minute relay would be used where the starting period is long or the machine is started frequently.

In setting the thermal section of the relay there are two adjustments which allow the correct setting to be made. A rough adjustment by changing the turns ratio on the auxiliary current transformer and a fine adjustment by alteration of the fixed contact position.

The possible settings on the auxiliary CT are 80%, 90% or 100% and a setting corresponding to the ratio of full load current to line CT primary current should be chosen.

Examples

45 kW motor FL current = 84 A, CT ratio 100/1,

$$\frac{84}{100} = 0.84 \text{ set to } 80\%.$$

55 kW motor FL current = 98 A, CT ratio 100/1,

$$\frac{98}{100} = 0.98 \text{ set to } 100\%.$$

75 kW motor FL current = 136 A, CT ratio 150/1,

$$\frac{136}{150} = 0.91 \text{ set to } 90\%.$$

75 kW motor FL current = 136 A, CT ratio 200/1,

$$\frac{136}{200} = 0.68 \text{ set to } 80\%.$$

The auxiliary CT, which incidentally utilises the magnetic circuit of the instantaneous overcurrent relay as a core, changes the overall ratio of the current transformer circuit. The 150/1 CT with the auxiliary CT on a 90% tap gives an overall ratio of

$$0.9 \times 150/1 = 135/1.$$

The adjustment of the fixed contact depends on the duty of the motor. If the motor load is fairly constant, say a fan or a pump, then the contact can be set fairly close to the full-load current value, say at 110%. If the load fluctuates, for example a conveyor, then a wider setting may be needed, a setting of 115% or even more.

This does not mean the fixed contact will be set at 110% or 115% although in most cases it will be fairly close to these values.

The contact should be set, for a 110% setting, to

$$110\% \times \frac{\text{FL current}}{\text{CT primary current} \times \text{aux. CT tap}},$$

in the case of the 37 kW motors

$$110 \times \frac{84}{100 \times 80} = 1.16 \text{ set, to } 116\%,$$

or the other motors

$$110 \times \frac{98}{100 \times 100} = 1.08, \text{ set to } 108\%,$$

$$110 \times \frac{136}{150 \times 90} = 1.11, \text{ set to } 111\%,$$

$$110 \times \frac{136}{200 \times 80} = 0.94, \text{ set to } 94\%,$$

or for a 115% setting

$$115 \times \frac{136}{200 \times 80} = 0.98, \text{ set to } 98\%.$$

The setting of the instantaneous overcurrent relay must be about $1\frac{1}{2}$ times the motor starting current in order to avoid operation during initial peak of the starting current which can be more than twice the steady short-circuit current but has a duration of less than one cycle.

It would be more correct to say that the relay setting should be about $1\frac{1}{2}$ times the motor short-circuit current. This is the same as starting current in the case of direct-on-line motors but not when the motor is started by a method which limits the starting current. The fact is that an induction motor will contribute current to a system fault at a level equal to the short-circuit current. This is the initial current at the moment of fault but quickly dies away.

Therefore even though the current has been limited during starting to, say, twice full-load current an external fault will cause a current of 6 to 8 times full-load current to flow from the motor. The instantaneous overcurrent relay must be set so that it does not operate under these circumstances. The actual setting is the times full-load figure on the scale multiplied by the tap setting on the auxiliary CT.

The success of this type of relay is undoubtedly due to its simplicity in setting and the ability to check its performance whilst in service. The moving contact arrangement carries a pointer which indicates the percentage load. From observation of the panel ammeter and a knowledge of the overall CT ratio the correct operation of the protection can be verified. The overall CT ratio is the line CT ratio, which can usually be deduced from the ammeter range, multiplied by the auxiliary CT tap which is indicated on the relay nameplate.

For example, the 45 kW motor, CT ratio 100/1, auxiliary CT tap 80%. If the panel ammeter was indicating 68 A, on the percentage load scale

$$\frac{68}{100 \times 0.8} \times 100\% = 85\%$$

should be indicated.

There are a number of electronic relays which protect the motor in the same way as the thermal relay which are capable of matching the motor characteristic more accurately. In addition to adjustments for current level the operating time can be adjusted as well as the settings for unbalanced current and earth fault.

The type of protection described would only be applied where the motor is supplied via a circuit-breaker. In many cases the switching of the motor is by a contactor which, although capable of making fault current, cannot interrupt fault current. Therefore fuses are used to clear the fault instead of the instantaneous overcurrent relays. The earth-fault relay is used as it will detect low-level faults and trip the

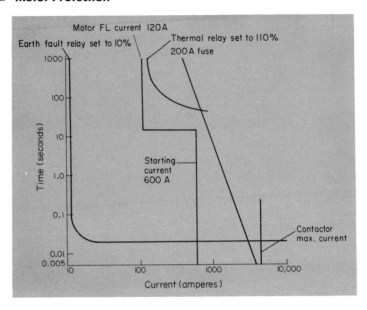

FIG. 9.4   COMPOSITE CHARACTERISTIC. THERMAL RELAY AND 200 A FUSE

contactor but for high-fault levels the fuse would operate faster than the earth-fault relay. It may be that a time delay has to be introduced into the earth-fault relay circuit to ensure that the fuse operates faster. Figure 9.4 shows a typical composite characteristic where a 160 A fuse is used in conjunction with a relay thermal element and earth-fault relay to protect a 550 kW motor with a full-load current of 120 A. From 132 A, 110% full load current to 700 A the motor is protected by the thermal relay which would trip the contactor. Above 700 A the 200 A fuse would clear the fault. Similarly earth fault from 12 A to 3000 A would be cleared via the contactor and above 3000 A by the fuse. The contactor will never be called upon to break a fault current beyond its capability.

DIFFERENTIAL PROTECTION

Differential protection on a phase by phase basis is shown in Fig. 9.5. This type of protection is eminently suitable and will detect faults on practically the whole of the winding but is generally only used on large motors. The leads between the current transformers in the motor neutral terminal box and the relay, which is associated with the switchgear, may be long and therefore could have a high resistance.

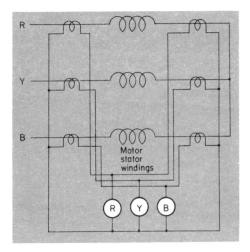

FIG. 9.5   DIFFERENTIAL PROTECTION

However, as the most onerous condition under which stability is required is the motor starting current a fairly low relay setting, and reasonably small current transformers, can be used.

LOSS OF SUPPLY

When the supply is removed from an induction motor its back e.m.f. will decay exponentially and virtually disappear in a few seconds. During that time there will also be a slight decrease in speed so that the phase of the back e.m.f. moves away from the position which it occupied before the removal of the supply. The result is that the locus of the back e.m.f. traces a spiral as shown in Fig. 9.6. The figures are arbitrary and are not meant to represent any particular motor.

If the voltage was restored before 0.4 s, then the voltage applied to the motor would be less than system voltage because of the back e.m.f. and the current would be less than short-circuit current. After 0.4 s, the voltage between the applied voltage and the back e.m.f. is greater than the applied voltage and the short-circuit current would be correspondingly greater. If the voltage was restored after 0.8 s, the short-circuit current would be $1\frac{1}{2}$ times normal. This means that the mechanical forces exerted on the rotor would be over twice the normal starting forces and could cause damage to the rotor structure.

For this reason undervoltage relays are used on large machines to ensure that the machines are disconnected if the loss of voltage

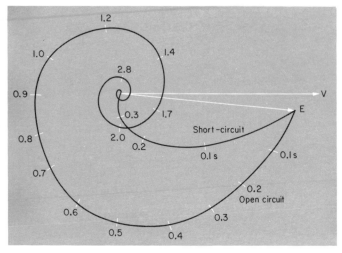

FIG. 9.6   LOCI OF BACK E.M.F. OF MOTOR DURING LOSS OF SUPPLY.

exceeds, say, 0.3 s. The relay used is either an attracted armature relay with a time-delay relay or an induction relay.

During a system fault there is a loss of supply to all motors connected to the system until the fault is cleared by unit protection. The loss of supply will be of the order of 120 to 250 ms, the protection-operating time plus the circuit-breaker opening time. Even if the fault persisted for a longer time there is not much danger of the high short-circuit current. This is because the motor will be contributing current to the fault and consequently the decay of the back e.m.f. is far more rapid. It will in all probability have disappeared in less than 0.5 s.

SYNCHRONOUS MOTORS

The protection for synchronous motors is the same as that for induction motors but with the addition of a relay to detect loss of synchronism and loss of supply.

For loss of synchronism an out-of-step relay is applied to motors which could be subjected to sudden overloads. The motor could pull out of step because of an increase in mechanical load or if there is a reduction in supply voltage. When pole slipping occurs the stator current increases and the power factor changes to a very low value and it is this condition that the out-of-step relay detects and trips out the motor during the first slip cycle.

The relay coil is connected to a bridge circuit which compares the current from one phase with the voltage from the other two phases. The relay is energised by the voltage and is in the operated position at

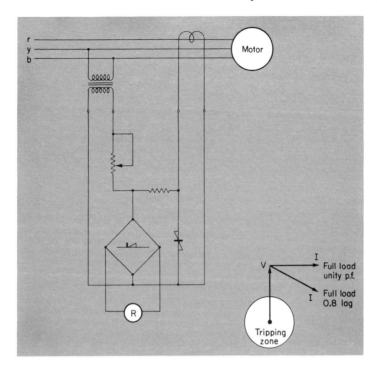

FIG. 9.7   OUT-OF-STEP PROTECTION. TRIPPING OCCURS WHEN THE VECTOR SUM
OF V AND I LIE WITHIN THE TRIPPING ZONE

all times when the current is zero or under normal load condition. When the motor pulls out of step the current is such that it subtracts from the current produced by the voltage to such an extent that the net current falls below relay drop-off level. A non-linear resistor increases the overall tripping area of the characteristic to ensure correct operation. This type of relay will also detect the loss of field.

If the supply to a synchronous motor is interrupted for more than, say, 0.3 s, then there is a danger that if the supply is restored the motor may be out of step and therefore an undervoltage relay is required to trip the machine. This relay will also prevent starting and running under abnormally low voltage conditions. Other protection devices are underpower or reverse power relays which are usually induction relays and are identical except that the former closes its contacts when the forward power is less than, say, 3% and the latter closes its contacts when the reverse power exceeds 3%. The reverse power relay should always be preferred as it is more stable to momentary swings of power but it depends for operation on the protected motor generating to other loads connected to the same busbar. If there are no other loads then an underpower relay must be used.

Chapter 10

Generator Protection

The a.c. generator needs protection against a number of conditions some of which require immediate disconnection and some that may be allowed to continue for some time. In broad terms the former are connected with insulation failure whilst the latter are generally associated with unsatisfactory operating conditions.

Of all the items of equipment which make up a power system the generator is unique in that it is usually installed in an attended station and is therefore subject to more or less constant observation. The point here is that some of the unsatisfactory operating conditions could be dealt with by an operator whereas if the generator was not attended tripping would be the only course of action.

INSULATION FAILURE

Stator faults are caused by the breakdown of the insulation between the armature conductor and earth; between conductors of different phases or between conductors of the same phase.

The most likely place for an earth fault to occur is in the stator slots. Arcing will probably occur resulting in the burning of the iron at the point of fault and welding the laminations together. Replacement of the faulty conductor may not be very difficult but the damage to the core cannot be ignored as the fused laminations could give rise to local heating. In severe cases it may be necessary to dismantle and rebuild the core which is a lengthy and costly process.

To reduce the possibility of damage earth-fault current is usually limited by earthing the generator neutral point via a resistor, reactor or transformer. Practice varies as to the value to which the current is limited. From rated current in some cases to very low values in others.

EARTHING BY RESISTOR

Earthing by means of reactors is uncommon and earthing by transformer is usually limited to large machines. In an industrial system the generator, which is usually directly connected to the power system without a transformer, is earthed by a resistor which has a fairly low value. The earth-fault current is usually limited to between 50% and 200% of the rated current.

In cases where the generator is connected to the distribution system via a generator transformer a resistor designed to allow an earth-fault current of about 300 A is used irrespective of generator rating.

EARTHING BY TRANSFORMER

The other approach to earthing is to limit the current to a level where burning does not readily occur. This level is said to be 5 A. To achieve this high-impedance transformers have been used. Initially voltage transformers were used operating at a fairly low flux density but overvoltage problems arising from the capacitance of the stator windings has resulted in the general use of distribution transformers. The secondary winding is loaded with a resistor so that under earth-fault conditions a maximum of 5 A will flow.

Phase-to-phase faults are far less likely than earth faults and, as they are easily detected, damage caused can be limited by rapid disconnection. On the other hand, interturn faults, which are also uncommon, are very difficult to detect and are generally only detected and cleared when they have developed into an earth fault.

STATOR PROTECTION

Differential protection using high-impedance relays is usual for stator protection and is applied on a phase-by-phase basis. As the leads between the two sets of current transformers may be long the resistance will be fairly high but as the maximum through-fault current will be less than 10 times full-load current a reasonably low voltage setting can be applied. This means that the CT magnetising current will be low and therefore a low overall current setting can be expected.

The overall setting has a direct bearing on the amount of the generator winding which is protected. This can be calculated as follows:

Max. fault current—say 5 × CT rating.
Overall protection setting—say 6%.
Amount of winding protected

$$100\% - \frac{6\%}{5} = 98.8\%.$$

This would be for a phase–phase fault. For an earth fault where the current is limited to the full-load value only 94% of the winding would be protected. In fact slightly less as the full-load current of the generator is usually less than the CT rating.

If the required voltage setting was high because of, say, long CT leads or if the CT magnetising current was high then the overall current setting may be much higher than 5%. This means that the amount of generator winding protected is also reduced maybe to an unacceptable level for earth faults. In this case a biased differential relay would alleviate the position.

The use of a biased relay means that the relay-coil circuit impedance can be reduced to about a twentieth of the impedance of the relay coil in the unbiased scheme. This naturally reduces the voltage setting and the CT magnetising current at setting resulting in an overall setting of about 5%.

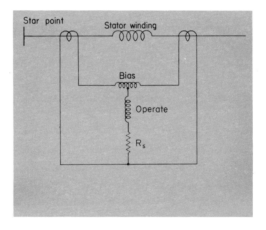

FIG. 10.1   BIASED DIFFERENTIAL PROTECTION. ONE PHASE ONLY SHOWN

The biased differential scheme is shown in Fig. 10.1 and the value of the stabilising resistor, R_S, can be calculated from

$$R_S = \frac{R_{CT} + R_L + \frac{1}{2}R_B}{B}$$

where B is the ratio of bias coil turns to operate coil turns and is known as the bias ratio and R_B is the resistance of the bias coil.

EARTH-FAULT PROTECTION

Where the maximum earth-fault current is restricted to a fraction of the generator rating earth-fault protection is essential to compliment the differential protection scheme.

This earth-fault protection frequently comprises an instantaneous relay having a setting of 10% to 20% and the IDMT relay with a setting of 5% to 10%. Both relays would be connected to a simple current transformer having a primary current rating equal to that of the earthing resistor. Earth faults will be detected in 90% to 95% of the generator winding even though the maximum earth-fault current may be as low as 5% of the generator rating.

Even where the main differential protection scheme is expected to provide adequate protection for earth faults an IDMT relay, connected to a current transformer in the generator neutral-earth connection, is used to provide back-up protection. Where the generator is directly connected to the power system, i.e. without a generator transformer, it provides back-up protection for the busbars and the whole system. In this case it should have a very long time delay and should be thought of as the last line of defence.

ROTOR EARTH-FAULT PROTECTION

The field system of a generator is not normally connected to earth and so an earth-fault does not cause any current to flow to earth and does not, therefore, constitute a dangerous condition. If a second earth-fault occurs a portion of the field winding may be short-circuited resulting in an unbalanced magnetic pull on the rotor. This force can cause excessive pressure on the bearings and consequent failure or even displacement of the rotor sufficient to cause fouling of the stator. The overheating in the rotor can cause deformation of the winding which could lead to the development of short-circuits.

Two main methods are used for detecting earth-faults in the rotor circuit. In the first method a high-resistance potentiometer is connected across the rotor circuit the centre point of which is connected to earth through a sensitive relay (see Fig. 10.2). The relay will respond to earth faults occurring over most of the rotor circuit.

There is, however, a blind spot at the centre point of the field winding which is at the same potential as the mid-point of the potentiometer. This blind spot can be examined by arranging a tapping switch which when operated shifts the earth point from the

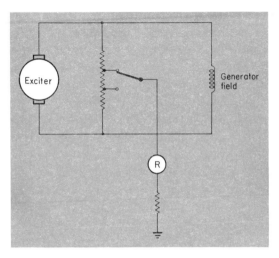

FIG. 10.2   ROTOR EARTH-FAULT DETECTION. POTENTIOMETER METHOD

mid-point of the potentiometer to a point a little to one side. The tapping switch can be mounted on the control panel and the check operation performed at suitable intervals.

An alternative method which avoids a blind spot consists of biasing the field circuit relative to earth by means of a simple transformer rectifier unit as shown in Fig. 10.3. This is connected between the positive bar of the field system and earth through a high-resistance relay. The positive bar of the field system is biased some 30 V negative to earth and therefore the remaining portions of the field circuit are proportionally more negative. A fault occurring at any point in the field system will apply a voltage to the relay which is sufficient to cause operation, the fault current being limited to a low value by the resistance of the relay. It is usual to connect the relay to give an alarm.

UNSATISFACTORY OPERATING CONDITIONS

These conditions in general do not require immediate disconnection and, it could be argued that, in an attended station the operator could take the necessary action to remove the condition. Undoubtedly this is possible in some cases but on no account should protection be omitted on this basis.

Unbalanced Loading

Unbalanced loading of the generator phases results in the production of negative phase sequence (NPS) currents. These currents, which have a phase rotation in the opposite direction to the normal

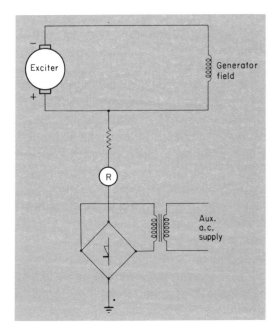

F<small>IG.</small> 10.3   R<small>OTOR</small> E<small>ARTH-FAULT</small> D<small>ETECTION.</small> N<small>EGATIVE</small> B<small>IASING</small> M<small>ETHOD</small>

phase rotation, produce a magnetic field which induces currents in the rotor at twice the system frequency. This causes considerable heating in the rotor and would cause damage if allowed to persist.

Each generator will have a negative phase-sequence rating which can exist continuously without damage, typically 0.15 p.u. of generator FL current, and an I^2t rating when the current exceeds the continuous value, typically $I^2t = 20$.

Where I is per unit NPS current and t is the time in seconds, e.g. the generator would carry a NPS of current 15% full-load current continuously and NPS current of 30% full-load current for a time of

$$0.3^2 t = 20 \quad t = \frac{20}{0.3^2} = 222 \text{ s.}$$

In fact the time would be longer than the calculated value as there would be some heat dissipation. An I^2t value assumes no heat dissipation and therefore the longer the time the more inaccurate the result. The result will be fairly accurate up to 2 minutes.

The actual negative phase-sequence current is difficult to determine from the ammeters measuring the load current in each phase. It is not greater than 65% of the unbalanced current.

Relays to detect the condition usually have an IDMT characteristic matched to the I^2t value. The relay is connected to a network which separates the positive and negative phase-sequence currents. The basis of the network is to produce a phase shift of 60° in some components of the phase currents such that when the phase rotation is positive, i.e. r, y, b, r, the net current in the relay is zero. When the phase rotation is negative, i.e. r, b, y, r, a proportion of the current flows in the relay. Any current which flows in the generator neutral is known as zero-sequence current and this must be eliminated if the network is to function correctly. Where the generator is connected to the system via a delta/star transformer any zero-sequence current means that there is a fault on the generator circuit and this will be cleared by earth-fault protection. If the generator is directly con- nected then zero sequence is eliminated by connecting in delta the current transformers which feed the NPS network. In this case the relay setting is related to the CT current × 1.73.

There is sometimes a reluctance to apply NPS protection as all generators will be subject to the same conditions and could lead to all generators tripping at the same time. An early warning of the condition can be provided by an instantaneous relay connected to the NPS network to provide an alarm after a short fixed time delay.

OVERCURRENT PROTECTION

An IDMT relay is generally used as back-up protection but the operation of this relay is complicated because of the current decre- ment in the generator during fault conditions. In some cases a setting is chosen, such that the relay will not operate for a system fault but will only respond when fault current is fed into the generator, in this way it only acts as a back-up to the main generator protection.

In most industrial installations the relay is required to act as back-up to the system protection and settings must be chosen to ensure positive operation.

The operation of IDMT relays under generator decrement con- ditions can be calculated by dividing the decrement curve into a number of zones of width, say 0.1 s. The mid-ordinate is the current level which is converted to a multiple of the relay setting and the time for full travel determined

$$t_1 = \frac{3}{\log M}.$$

Therefore in 0.1 s

$$\text{travel } x = \frac{0.1}{t_1} = \frac{0.1 \log M}{t_1}.$$

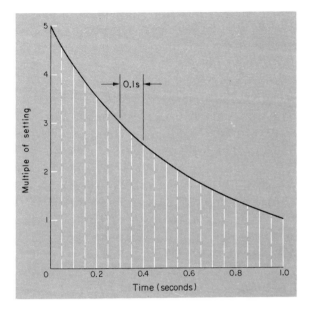

F<small>IG</small>. 10.4   G<small>ENERATOR</small> D<small>ECREMENT</small> C<small>URVE WITH</small> M<small>ID-ORDINATES FOR THE</small>
      C<small>ALCULATION OF</small> R<small>ELAY</small> R<small>ESPONSE</small>

To this is added the value of x calculated from the multiple of current at the next mid-ordinate, and so on until the required time is reached. The total travel time will be the time multiplier setting required to give relay operation in the required time.

Figure 10.4 shows the current decrement curve of a typical generator on no load prior to the fault. A current of 5 times relay setting at $t = 0$ decays to the level of relay setting after 1 s. The mid-ordinates of the 0.1 s strips are as in Table 10.1.

T<small>ABLE</small> 10.1

t	Mid-ord value of M	Travel $= \dfrac{0.1 \log M}{3}$	Total travel
0–0.1	4.6	0.022	0.022
0.1–0.2	3.8	0.019	0.041
0.2–0.3	3.3	0.017	0.058
0.3–0.4	2.8	0.015	0.073
0.4–0.5	2.3	0.012	0.085
0.5–0.6	2.0	0.01	0.095
0.6–0.7	1.7	0.008	0.103
0.7–0.8	1.5	0.006	0.109
0.8–0.9	1.3	0.004	0.113

Below 1.3 times setting operation cannot be guaranteed.

Table 10.1 shows that with a time multiplier setting of 0.1 the relay will have travelled that distance in 0.6 to 0.7 s.

The difficulty in application arises from the variation in the current decrement depending on generator conditions prior to the fault. From a no-load condition the current will decay to less than full-load current whereas from the full-load condition the final current will be greater than full-load current because the field current is higher. The former case will be modified if there is a voltage regulator as this will attempt to boost the field with a consequent increase in final current. This would have a large effect on the relay and therefore a normal IDMT relay is generally unsatisfactory. However, this method can be used to determine settings of feeder and transformer IDMT relays in finite busbar systems. For example, in off-shore installations or any location where the only supply is local generation. The multiples of setting current in this case will be much greater because the feeder and transformer rated current will only be a fraction of that of the generator. The higher multiples of setting means that the effect of the difference in generator decrement between no load and full load will be small.

It may be that the current will decay to a level where it is insufficient to cause the overcurrent relay to trip. In these circumstances it is necessary to provide a relay which not only responds to current but also to the level of voltage.

The principle of operation is that an IDMT relay with a setting much less than the full-load current of the generator has a feature added which increases the setting to above full-load current when full system voltage is present.

By this means the longer operating times, for discrimination with system protection when the fault is remote, will be attainable as the voltage is high. Close-up faults will remove the voltage restraint to enable the relay to operate in the relatively fast time appropriate to the lower setting.

The relays for this type of protection can be either voltage restrained, where the voltage element operates as a restraint on the same disc as the overcurrent element, or voltage controlled, where the setting of the overcurrent relay is changed by means of a voltage-operated attracted-armature relay.

OVERLOAD

Overload protection is not generally provided for continuously supervised machines but on large machines resistance thermometers or thermocouples are embedded in the stator winding. There is some possibility of overload in terms of MVA for, although the governors will restrict MW, the AVR may cause the machine to deliver a disproportionate share of the MVAr. In cases where overload protection is to be provided this would probably be of the thermal type with a characteristic to match the generator thermal capacity.

Overload and overcurrent relays should not be confused as they perform completely different functions. An overload relay operates in the hundreds to thousands of seconds range whereas an overcurrent relay operates in the one- to ten-second range.

FAILURE OF PRIME MOVER

In the event of a prime mover failure the generator continues to run but as a synchronous motor and this can cause a dangerous condition in the prime mover. In a steam turbine the turbulence of the steam in the turbine causes a temperature rise which can quickly reach serious proportions in pass-out sets. In condensing sets the temperature rise is not as fast and therefore less urgent action is needed. In engine-driven sets the loss of motive power is likely to be due to mechanical failure and the continued running of the set is likely to cause damage.

The machine, as a synchronous motor, will draw power from the system and it is this reverse power which is detected by the protection. The power required is usually small, about 10% in case of large turbo-alternators. The power factor depends on the excitation of the machine and can be quite low and either leading or lagging. This means that the reverse power relay must respond to a low value of power when the MVAr is high and consequently must be sensitive and have only a small phase-angle error.

A single-element relay is used because the power will be balanced in the three phases. It is used in conjunction with a time-delay relay to prevent operation during power swings and synchronising.

LOSS OF FIELD

Failure of the field system results in acceleration of the rotor to above synchronous speed where it continues to generate power as an induction generator the flux being provided by a large magnetising component drawn from the system. This condition can be tolerated

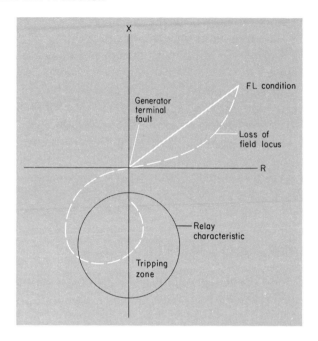

FIG. 10.5   DETECTION OF GENERATOR FIELD FAILURE

for a short time but clearly there will be increased heating of the rotor because of the slip-frequency currents which flow.

Loss of field can be detected by a simple undercurrent relay connected to a shunt in the field circuit. It must have a setting below minimum field current and a time delay if field forcing is used. A time-delay relay is also required as the undercurrent relay may respond to the slip-frequency circuit in the field circuit. This relay would have an instantaneous pick-up and a time-delayed drop off to maintain the circuit to the main time-delayed relay.

The more up-to-date method is to detect the loss of field on the a.c. side of the generator by comparison of the stator voltage and current. By either a relay measuring the reactive power (MVAr) which is being imported or by an impedance relay which has a characteristic as shown in Fig. 10.5. As can be seen under normal operation the apparent impedance, as measured by stator voltage and current, is well away from the tripping zone. When there is a loss of field the impedance vector moves to the operation zone.

OVERSPEED

The speed is very closely controlled by the governer and is held constant as the generator is in parallel with others in an interconnected system. If the circuit-breaker is tripped the set will begin to accelerate and although the governer is designed to prevent overspeed a further centrifugal switch is arranged to close the steam valve.

There is still a risk, however, that the steam valve will not close completely and even a small gap can cause overspeed and so where urgent tripping is not required it is usual to lower the electrical output to about 1% before tripping the circuit-breaker. A sensitive underpower relay is used to detect when this value is reached.

OVERVOLTAGE

Voltage is generally controlled by a high-speed voltage regulator and therefore overvoltages should not occur and overvoltage protection is not generally provided for continuously supervised machines. On unattended machines an instantaneous relay set at, say, 150% is used to cater for defective operation of the voltage regulator.

PROTECTION OF GENERATOR/TRANSFORMER UNITS

Where a generator is connected to the power system by means of a generator transformer it is usual to protect the generator and transformer as a single unit using biased differential protection.

The current transformer balance is produced in terms of both phase and magnitude, i.e. in the arrangement shown in Fig. 10.6 there is an overall phase change of 30° which is corrected by connecting a set of auxiliary current transformers in delta. Because of the difference in current transformer ratios the settings of the generator transformer protection has to be somewhat higher than the settings of the generator protection. Because of this the generator is sometimes protected separately but is also included within the zone of the generator-transformer protection as an extra insurance. The transformer is connected directly to the generator and so no harmonic restraint circuit is required in the protection as no switching can occur. There is a low level of magnetising inrush current following a fault when the voltage is restored from being depressed but this is usually insufficient to unbalance the protection. Figure 10.6 shows a complete protection system for a generator.

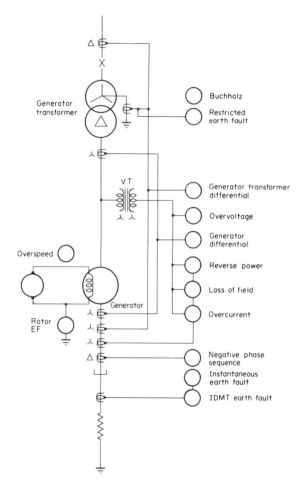

F<small>IG</small>. 10.6   G<small>ENERATOR</small> P<small>ROTECTION</small>

UNIT TRANSFORMERS

A unit transformer is frequently teed off the generator transformer connection and an extra set of current transformers in this tee are required to balance the differential protection, separate differential protection is required for the unit transformer as its rating is low compared to that of the main transformer and the main protection would not deal with unit transformer faults adequately. The unit transformer protection would have current transformers matched to

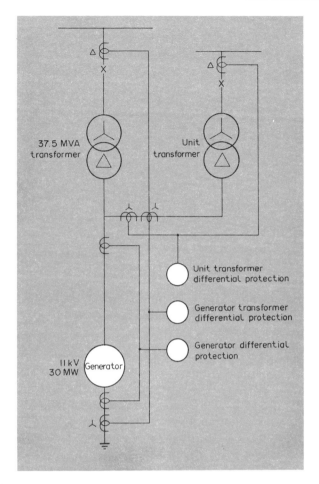

FIG. 10.7   DIFFERENTIAL PROTECTION OF GENERATOR, GENERATOR
              TRANSFORMER AND UNIT TRANSFORMER

its rating thus producing a current setting commensurate with the transformer size. Figure 10.7 shows the arrangement.

In addition the unit transformer will have its own overcurrent and earth-fault protection.

Chapter 11

Control Circuits

The d.c. circuits associated with protection are, in the main, tripping circuits which are an essential link between the protection and the circuit-breaker. The importance of this link cannot be stressed too strongly because however much thought there is in the protection design or however much protection is provided they are of no avail if the trip circuit fails to function.

In an automatic control circuit equipment is in continuous operation and any fault is discovered quickly. On the other hand, a tripping circuit may operate only once, and then at a time where conditions are abnormal, and therefore there must be no risk of failure.

The vital requirement, therefore, is reliability which means that the performance of the circuit must be completely predictable. Simplicity in design allows this objective to be achieved.

Figure 11.1 shows the simplest tripping circuit. The protection contacts energise the circuit-breaker trip coil via a normally open auxiliary contact which opens when the circuit-breaker opens to break the trip-coil circuit. This is necessary because the trip coil is short-time rated and because most protection relay contacts are incapable of breaking this highly inductive circuit.

It will be noted that in this case there are links in both the positive and negative sides of the circuit. The protection would be by a single fuse, common to all tripping circuits in the battery positive connection. It is quite easy to monitor that this fuse is intact.

An alternative philosophy is to fuse the positive side of each trip circuit on the basis that any wiring fault will only render one trip circuit inoperative. This approach is a particularly sound philosophy if used in conjunction with trip circuit supervision relays.

Where the whole of the tripping circuit between the links, or fuse and link, is contained within the switchgear cubicle the possibility of a short-circuit is small and the use of a link in the positive connection will provide more security. Where the protection, or other tripping contacts, are located some distance from the switchgear, for example, Buchholz relays, cooling water or oil pump failure devices, etc.,

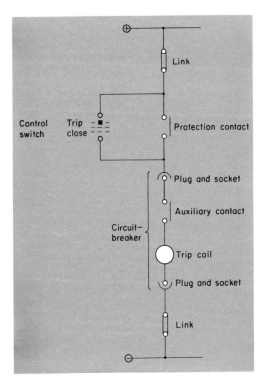

FIG. 11.1   BASIC TRIPPING CIRCUIT

short-circuits are much more likely and these circuits should be fused. In addition it would be good practice to use trip circuit supervision.

The circuit can be supervised by the simple expedient of connecting a relay coil across the remote contacts. This relay should have a pick-up current which is a small fraction of the normal trip-coil current. Contacts on the relay would be arranged to give an alarm when the circuit was broken and therefore it is necessary to inhibit this alarm when the circuit-breaker is opened as the circuit-breaker auxiliary switch would open the circuit. There are two ways either by interrupting the alarm circuit by a second circuit-breaker auxiliary contact or by connecting a resistor across the auxiliary contact in the trip-coil circuit to ensure continuity when the circuit-breaker is open as shown in Fig. 11.2. In this case some method of inhibiting the alarm when the circuit-breaker is isolated is required.

Many protection relays have auxiliary contactors which are used either to increase the number of relay contacts or to reinforce the protection contact.

Auxiliary contactors are usually attracted-armature relays and, because of their snap action, the contact force produced is considerable. In other types of relay this force is somewhat less and difficulties

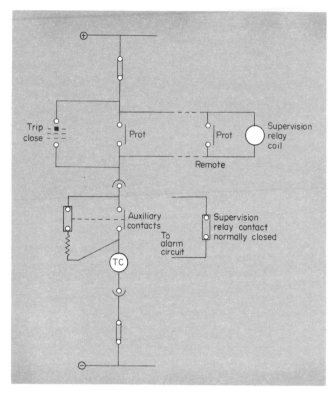

FIG. 11.2   TRIP CIRCUIT SUPERVISION

could arise if they were required to make more than one contact. Therefore in circuits where more than one output contact is required an auxiliary contractor with two, four or more contacts is used. The contacts do not need to be the same mode, i.e. normally open, as the protection contact but can be a combination of normally open and normally closed contacts to suit the requirement of the tripping, alarm or control circuit with which they operate.

In some cases the protection relay contact is incapable of carrying the trip-coil current and an auxiliary contactor is needed to reinforce this contact. The contactor can be used in either a series or shunt mode (Fig. 11.3). Although the protection-relay contact may be unable to carry the trip-coil current there is no difficulty in making the current as in the inductive circuit the current increases relatively slowly and before the protection relay contact rating is exceeded it is reinforced by the auxiliary contact.

In some installations auxiliary relays are used to indicate the cause of tripping particularly where the device causing the tripping is a simple contact and is not equipped with an indicator. Figure 11.4 shows a typical scheme for a large motor. As there is only a simple

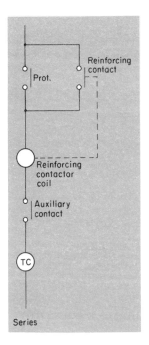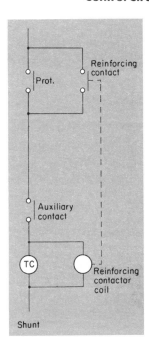

FIG. 11.3   AUXILIARY CONTACTOR USED TO REINFORCE PROTECTION RELAY
CONTACTS

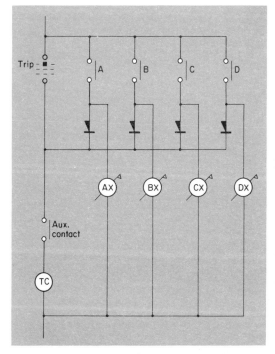

FIG. 11.4   THE USE OF A
SINGLE CONTACT
FOR TRIPPING
AND INDICATION

A	Overcurrent Relay Contact
B	Bearing Oil Pressure Switch
C	Winding Thermostat
D	Bearing Vibration Monitor
AX, BX, etc	Flag Indicator Relays

contact available it must perform the dual function of tripping and operation of the indicator relay. The half-wave rectifier, which must be rated to carry trip-coil current, allows this.

BATTERIES

The most vital part of the tripping circuit is the battery and because of this it should be continuously monitored for earth faults and low voltage and be serviced frequently.

The usual mode of operation is with the battery unearthed, except for the earth-fault detection device and continuously trickle charged. It is a good plan to monitor the a.c. supply to the trickle charger. Batteries should have sufficient capacity to deal with the tripping of all the circuits to which it is connected following the closing of at least two circuit-breakers.

A typical 110 V tripping battery would consist of fifty-five cells and when fully charged would be at a voltage of 126 V. This should be borne in mind when connecting any coil which could carry current continuously.

POWER FACTOR CORRECTION

The principle of power factor correction is widely applied in situations where there is a penalty clause in the supply agreement but in addition to this obvious benefit there is also a reduction in cable and transformer copper losses and the possibility of increasing the power which can be distributed by the power system. For example, a cable carrying current at 0.9 power factor can carry 25% more power than the same cable carrying current at 0.7 power factor.

In many installations a suitable capacitor is installed and permanently connected in circuit irrespective of the load and load power factor. Other installations have capacitors connected to certain loads which are switched in when the loads are switched in. These methods are not making the best use of power factor correction capacitors as it is possible at times to overcorrect which results in a leading power factor. This increases the current and therefore the cable and transformer copper losses and reduces the power-carrying capacity of the system.

The most effective use is made when the capacitor is switched in by a relay which is monitoring the reactive power of the system. It may be switched into circuit as one unit or, even more effectively, as a number of steps.

An induction disc relay is used with a voltage and current coil the former connected to, say, R–Y volts and the latter carrying B current. This is a quadrature connection which means that the maximum relay torque is produced when the current lags or leads the voltage by 90°. The relay has two contacts designated "lead" and "lag" and a range of adjustment from about 2% to 16% with a tapped current coil to extend this range by multiplying factors of 2 and 5.

Tap	Multiplying factor	Range
1	1	2–16%
2	2	4–32%
3	5	10–80%

A system rated at 1 MVA which has a power factor of 0.7 is capable of delivering a load of 700 kW and the reactive power is about 700 kVAr. If a power factor correction capacitor of 450 kVAr is installed the load can be increased to 900 kW because of the reductions of the reactive power to 250 kVAr.

The relay controlling the switching in and out of circuit must have a setting less than 480 kVAr, otherwise the capacitor will not be switched out after being switched in, and greater than half of 480 kVAr to prevent hunting. The setting should be about two-thirds of the kVAr which is being switched, i.e. 320 kVAr. Figure 11.5 shows a diagram of the performance. As the load increases from a low value the kW and kVAr increase at the same rate until the reactive power reaches the 320 kVAr lagging which is the relay setting. The

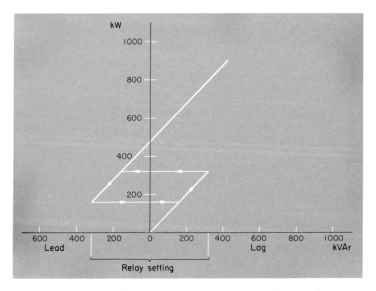

FIG. 11.5   POWER FACTOR CORRECTION BY SWITCHING A SINGLE CAPACITOR

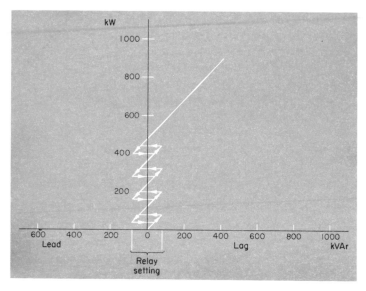

relay causes the capacitor to be switched in and the reactive power becomes 160 kVAr leading. A further increase in load reduces the kVAr to zero at a load of 480 kW and then increases the kVAr in the lagging directioin so that at 900 kW the power factor is 0.9. A reduction in load leaves the capacitor in circuit until a reactive power of 320 kVAr leading is reached.

Although the kVAr is contained within 320 kVAr leading and lagging the power factor is not very closely controlled and is poor at low levels of load. For more accurate control the capacitor could be switched into circuit in a number of steps as shown in Fig. 11.6. In this case the 480 kVAr is switched in four 120 kVAr steps. The relay setting being 80 kVAr leading and lagging.

As previously mentioned the relay is calibrated in percentage of nominal kVAr and the relationship between the setting and actual kVAr is established in the following manner.

Considering the system which is rated at 1 MVA at 415 V with a CT ratio of 1500/1. The power factor relay would be rated at 415 V and 1 A and therefore the normal rating of the relay is

$$100\% = \sqrt{3} \times 415 \times 1500 \times 10^{-3} = 1078 \text{ kVAr}.$$

Therefore a setting of 320 kVAr in terms of the relay rating would be

$$\frac{320}{1078} \times 100\% = 29.7\%.$$

Set to 30% (plug 2, 15%),

and a setting of 80 kVA would be $\dfrac{80}{1078} \times 100\% = 7.5\%$.

Set to 7.5% (plug 1, 7.5%).

In a typical installation where the 415 V supply is by means of, say, two 11,000/415 V transformers the power-factor correction capacitor would be connected on the 415 V side. This would result in less current, and therefore less copper losses, in the transformers. The measurement of the reactive power, however, should be made on the 11 kV side in order to include transformer magnetising current.

There would be no difficulty in calculating the setting as kVA and kVAr are used. For example, if a setting of 320 kVAr is required and the CT ratio is 75/1 and the VT ratio 11,000/110. The normal rating of the relay is

$$100\% = \sqrt{3} \times 11{,}000 \times 75 = 1429 \text{ kVA}.$$

Therefore the setting is $\dfrac{320}{1429} = 22.4\%$.

If the installation consisted of two power factor correction capacitors, one on each 415 V busbar section, two relays would be used each associated with one transformer. If, however, the power factor correction capacitor was only one unit a single relay would be used fed by an auxiliary current transformer which would summate the CT output from both transformers.

The auxiliary transformer would have two primary windings and a single secondary winding connected to the relay and would be designated typically $1 + 1/1$. The inclusion of the auxiliary transformer would double the normal rating. There is no limit to the number of primary windings. A $1 + 1 + 1/1$ auxiliary transformer would be required if the current of three distribution transformers was to be summated. The only requirement is that the main current transformers must all have the same ratio.

If in the above installation there was no 11 kV voltage transformer then the relay voltage coil could be connected to the 415 V supply whilst the current coil was still connected to the 11 kV side current transformers. In most cases the transformer would be delta/star and therefore there is a phase change of 30° between primary and secondary sides which must be taken into account to ensure that the relay will measure kVAr. As can be seen from Fig. 11.7 this is achieved by connecting the voltage coil between line and neutral.

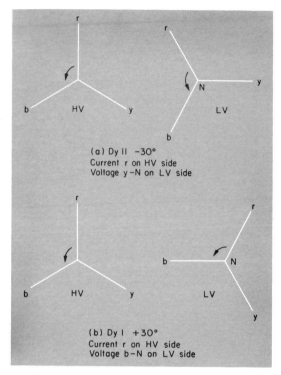

FIG. 11.7   RELAY CONNECTION FOR MEASUREMENT OF kVAr

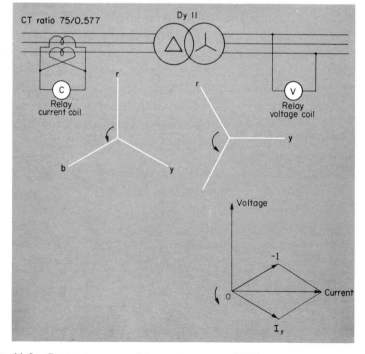

FIG. 11.8   CONNECTION FOR MEASUREMENT OF KVAr

The voltage coil would have to be rated for 240 V but the calculation of normal rating would be as before,

i.e. $100\% = \sqrt{3} \times 11{,}000 \times 75 = 1429\,\text{kVAr}$.

If the neutral is not available then an artificial neutral can be produced by two coils which have impedances similar to the relay voltage coil. Alternatively the 90° phase shift can be produced by the connections shown in Fig. 11.8. In this case the cross-connected current transformers would have a secondary current rating of 0.577 A so that the combination would produce 1 A.

Chapter 12

Testing

The importance of maintenance and testing of protective gear will be appreciated if its role in the power system is considered. It is different from all other equipment in that it is operative for a very small proportion of the time. It is therefore most important that when operation is required it will function correctly.

To ensure this, regular maintenance and testing of the relay and its associated equipment is required. It is not possible to specify the frequency of testing—this depends on the location and the import-ance of the equipment. An important piece of equipment with protection mounted in a location where conditions are poor would require some attention every 12 months, or maybe more often in the light of experience, whereas a less important unit with protection in a good location would require testing every 4 years. There is no hard-and-fast rule—a case for judgement and common sense and experience.

WORKS TESTS

To appreciate the aims of site testing it is necessary to consider the tests to which a relay is subjected in the manufacturers' works.

During the development stage of a relay, or protection scheme, many tests are performed to achieve the desired characteristics and performance. When development is completed the equipment is subjected to a type test. This is in two parts. The tests which are common to every relay, impact, vibration, insulation tests, etc., and the tests which are particular to the relay to prove its characteristics, speed of operation, stability level, CT requirements, etc.

It is information derived from these tests which will ultimately be used in leaflets describing the relay and specifying its performance. These type tests would be performed on only a few of the relays and none of these would be supplied to customers. Relays which are supplied against customers' orders are subjected to a series of tests to prove that their characteristics conform, within limits, to the specification and that it will perform in the manner described in the technical literature describing the relay. The tests required are enumerated in a test specification which usually culminates in a pressure test to check the insulation.

TESTS ON SITE

The relays are delivered to the site either mounted on a switchgear panel or as "loose" relays for mounting on the panel on site. In the latter case a check should be made before mounting to confirm that the relay has not been damaged in transit.

Tests conducted on the relays when they are installed in their final location are to prove that the connections to the relay are correct and that there is no damage or foreign matter introduced into the relay during installation and, in the case of relays delivered already mounted on the switchgear panel, in transit.

It is desirable to have a wiring diagram of the equipment which is to be tested as this will reduce considerably the time to perform the tests. All the wiring should have ferrule numbers and it is a simple matter to relate these to the wiring diagram.

Another aid is the use of standard wiring numbers which are used by many switchgear manufacturers. These follow the recommendations of BS 158:1961 even though this particular British Standard Specification is now withdrawn. From a knowledge of the nomenclature the function of much of the panel wiring can be deduced without reference to a wiring diagram.

The ferruling consists of a letter which refers to a function and a number which in the case of CT and VT circuits refers to a phase. A, B, C and D are CT circuits, A, B and C are for differential, busbar and overcurrent circuits respectively whilst D is for metering circuits. E is associated with VT circuits. The numbers 10 to 29, 30 to 49 and 50 to 69 refer to R-, Y- and B-phases respectively whilst 70 to 89 are for neutral and residual circuits whilst 90 is for connections made directly to earth. Figures 12.8 and 12.9 are examples of the use of these numbers. In addition the letter K is used for tripping circuits and L for indication and alarm circuits. The usual practice is to use odd numbers for connections on the positive side of the supply and even numbers for connections on the negative side.

When tests are conducted it is most important to record the results in a clear and legible manner. This not only allows the results to be examined later, but provides a permanent record of the state of the equipment at that time and provides a basis for comparison for future tests.

COMMISSIONING TESTS

During commissioning a comprehensive series of tests are required to check the whole installation from the current transformers to the tripping circuit.

The tests can be divided into five parts:

CT Polarity Check
CT Magnetising Curves
Relay Characteristic Check
Insulation Tests
Tripping Circuit Check.

CT POLARITY CHECK

In many protection schemes the relative polarity between current transformers is important and therefore tests must be carried out to ensure that they are correctly connected. Figure 12.1 shows the diagram of a current transformer with the current flow convention which is when primary current flows from P1 to P2, secondary current flows from S1 to S2 in the external circuit connected to the current transformers. A simple way of checking the relative polarities is by

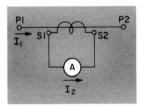

FIG. 12.1   CURRENT TRANSFORMER SHOWING CONVENTIONAL CURRENT FLOW

the flick test which uses a battery to send a pulse of current through the current transformer as shown in Fig. 12.2. If a d.c. current is passed through the CT from P1 to P2 then there will be a momentary deflection of a voltmeter connected across the secondary winding terminal S1 being momentarily positive. When the current is removed, terminal S2 will become momentarily positive. The usual method is, however, by primary injection.

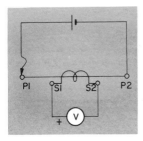

FIG. 12.2   CHECKING CT POLARITY BY FLICK TEST

Primary Injection

Primary-injection testing involves the passing of heavy currents through the current transformers to establish firstly the ratio and then the relative polarity. A short-circuit is placed as near as possible to the current transformers and current injected. The usual method of injection is into the switchgear feeder orifices by means of expandable rods and placing the short-circuit in the cable connecting box. If the latter is compound filled or if the connections are sleeved then the short-circuit would have to be placed in the CT chamber itself. The arrangement is shown diagrammatically in Fig. 12.3.

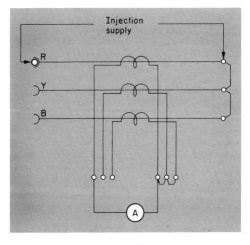

FIG. 12.3   RATIO CHECK ON RED PHASE CURRENT TRANSFORMER

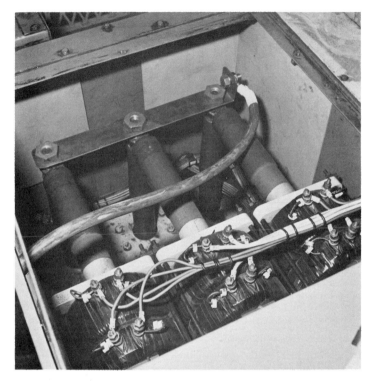

PLATE 12.1   PRIMARY INJECTION TESTING—A SHORTING LINK CONNECTED
IN THE CT CHAMBER (British Steel Corporation)

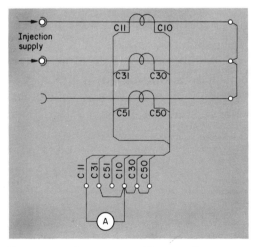

FIG. 12.4   RATIO CHECK WITH INJECTION AT FEEDER ORIFICES

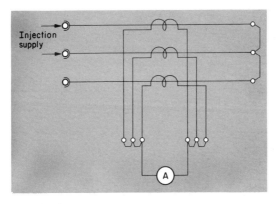

Fɪɢ. 12.5   Pᴏʟᴀʀɪᴛʏ Cʜᴇᴄᴋ. Rᴇᴅ ᴀɴᴅ Yᴇʟʟᴏᴡ Cᴜʀʀᴇɴᴛ Tʀᴀɴsꜰᴏʀᴍᴇʀs

After making the heavy current connections to inject current in, say, the red phase an ammeter is connected across the CT secondary and a short-circuit to any other CT which will be subjected to primary circuit during the test. The secondary connections are usually made at the main connecting block.

As an alternative the current could be injected via two feeder orifices, say the red and the yellow phases. In this case ammeters could be connected to both red and yellow current transformers or one CT short-circuited whilst the other CT ratio was checked. This arrangement using wiring numbers is shown in Fig. 12.4.

Current is injected via the two feeder orifices to check that the polarity of the current transformer is correct. The current which will pass through two current transformers in opposite directions and will produce a secondary current in each secondary winding. If the polarity of the current transformers is correct this current will circulate around the two secondary windings and very little current will flow in the ammeter which is connected as shown in Fig. 12.5. If the polarity is incorrect then the sum of the currents in the secondary windings will pass through the ammeter.

Should an incorrect polarity be indicated then the injection through a different phase will reveal the incorrect current transformer, e.g. if an incorrect polarity is shown when the injection is via R and Y and also when the injection is via Y and B then the Y current transformer is incorrect. If the R–Y injection shows an incorrect polarity and a Y–B injection does not the R current transformer is faulty.

In all cases, whether a correct or incorrect polarity is shown in the two tests, the third test should be performed as a double check.

PLATE 12.2   PRIMARY INJECTION TESTING. Inserting plugs into the switchgear orifices. After insertion a knurled-headed bolt is used to open the end of the plug (British Steel Corporation)

Having checked the current transformer ratios and that the relative polarity is correct the main function of primary-injection testing has been fulfilled and further testing can be performed by simulating the current transformer secondary currect by injecting a current at the CT secondary terminals. This current will, of course, be much lower than the primary-injection current and the equipment smaller. It is not possible to test the relays completely by the primary-injection method owing to the difficulty in producing high multiples of setting current. Therefore further primary-injection testing is confined to injecting sufficient current to produce movement of the relay.

Primary-injection tests are performed only when the equipment is being commissioned or if for any reason one or more current transformers are changed.

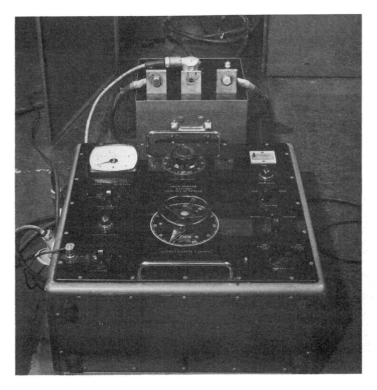

PLATE 12.3   PRIMARY INJECTION TEST SET. Control unit and low voltage injection transformer (British Steel Corporation)

CT MAGNETISING CHARACTERISTIC CURVES

This test is conducted on all current transformers and is intended to prove that they are suitable for the protection with which they are associated by determining the knee-point voltage, i.e. the voltage at which saturation starts. It is also intended to show that all current transformers in a group are similar and to test for open-circuited secondary windings or short-circuited turns. A variable voltage supply is connected across the secondary terminals of the current transformer and the current measured at different voltages. The circuit is as shown in Fig. 12.6. Note that the voltmeter is connected so that the ammeter does not read voltmeter current which in some cases could be of the same order in the CT magnetising current.

It is useful to establish roughly the voltage at which saturation starts

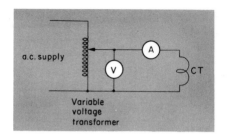

FIG. 12.6   CIRCUIT TO DETERMINE CT MAGNETISING CHARACTERISTIC

by increasing the voltage until there is a large increase in current for a small change in voltage. It can then be decided at what values to take readings to give sufficient points to plot a clear graph—too many points would be tedious.

As an example take the following tests on a set of three current transformers.

The test equipment is connected to the R-phase current transformer. The voltage is increased steadily until there is a rapid increase in current. In this case the increase was from about 0.05 A to 0.1 A when the voltage increased from about 100 to 120 V. Ten readings would seem to be a reasonable number between 0 and 120 V, say 20 V, steps initially and smaller steps when saturation starts. The Y- and B-phases are checked in turn.

Voltage	Current		
	R	Y	B
0	0	0	0
20	0.01	0.01	0.011
40	0.016	0.016	0.017
60	0.023	0.024	0.025
80	0.031	0.031	0.032
90	0.037	0.037	0.039
100	0.046	0.047	0.049
110	0.065	0.066	0.069
120	0.105	0.107	0.111

The knee-point voltage, i.e. the voltage at which an increase of 10% will result in a 50% increase in magnetising current, is about 100 V, the point in our preliminary test where the current started to increase rapidly.

As a rough check on this

$$100 \text{ V} \qquad\qquad I_e = \quad 0.046$$
$$100 + 10\% = 110 \text{ V} \qquad I_e = \quad 0.065$$

$$\text{increase} \quad \frac{0.065 - 0.046}{0.046} = 0.41$$

a 41% increase.

$$110 \text{ V} \qquad\qquad I_e = \quad 0.065$$
$$100 + 10\% = 121 \text{ V}$$

$$\text{at } 120 \text{ V} \quad I_e = \quad 0.105$$

$$\text{increase} \quad \frac{0.105 - 0.065}{0.065} = 0.62$$

a 62% increase.

Hence the knee-point voltage is between 100 and 110 V. If a more accurate result is required curves may be plotted as shown in Fig. 12.7. The knee point is 104 V for the R current transformer. Following these tests the resistance of the CT secondary winding is checked.

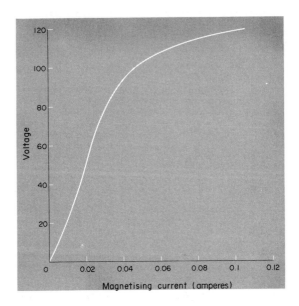

FIG. 12.7 CT MAGNETISING CHARACTERISTIC

This is best done by a bridge or by d.c. voltage and current measurement but a multimeter reading is better than nothing. All results must, of course, be recorded.

RELAY CHARACTERISTIC CHECK

The method employed to test the relay and the CT secondary wiring is secondary injection which, as previously mentioned, is a current injected at the current transformer secondary terminals to simulate the CT output. This is the usual method used during routine maintenance tests or tests following a suspected malfunction of the equipment.

The actual tests would be designed to show any defect in the equipment and to record the performance for comparison to its performance during the subsequent tests. This means that the same tests must be performed each time and it is therefore necessary to have a test sheet to ensure this.

It is not possible, and indeed not desirable, to provide a single test sheet which could be used on every type of equipment and therefore a test sheet should be prepared for each type of relay. The objection to a "universal" test sheet is that many of the spaces left for results would not be relevant and spaces that required a result would be overlooked. It is more satisfactory to have a test sheet in which every space requires a result. In this case the tests required are quite clear and the omission of a test unlikely.

As an alternative to the above if the protection on the particular system is fairly standardised the test sheets could be prepared for each type of equipment. For example, a transformer protection panel could have a test sheet which covered overcurrent, balanced earth-fault and Buchholz.

Typical test sheets for the more common types of protection could be as shown in Table 12.1.

The test at $1.3\times$ setting produces a very low disc torque and is to check that the disc runs freely showing that the bearings, disc and disc shaft are in good order and that there are no foreign bodies in the magnet gaps. The operating time is not very significant unless it varies widely from 30 to 40 s. This is because a small error in current measurement at this level affects the time considerably.

The tests at 2 and 10 times setting are to check the characteristic curve. Note that the relay is only short-time rated at 10 times setting as its consumption is about 150 W and so the current should be removed as the relay operates.

TABLE 12.1. IDMT OVERCURRENT AND EARTH-FAULT RELAYS

Phase	Function	Sec at 100%, * 1.0				Relay setting			Sec at 2×	HS amps	No plug
		1.3×	2×	10×	Reset	Plug	TM	HS			
R	O.C.										
Y	O.C.										
B	O.C.										
E	E.F.										

* E.F. relay at 40%.

The resetting time is checked by releasing the disc from the fully operated position. The disc should be watched for erratic movement which would indicate bearing problems, etc. The time to reset, which should be about 12 s, confirms that the permanent magnet has not lost any of its magnetism. The test with the plug removed is to check that the shorting switch, which prevents the open-circuiting of the current transformers when the plug is removed, is in working order.

After checking the resetting time the settings which have been calculated for the particular relay are applied and the operating time at 2× setting is measured. If an instantaneous overcurrent relay (HS) is fitted then its operating current is checked.

Thermal Relays

To check the overload characteristic requires that the current in each phase is balanced and therefore it is usual to connect the three-phase elements in series. Figure 12.8 shows a typical wiring diagram for a thermal relay with the current transformers connected to the 100% tap. Figure 12.9 shows the modifications which need to be made in this case to prepare the relay for a secondary injection test. The earth-fault element and the yellow phase CT are short-circuited and the C71 connection between the red-phase and the earth-fault elements is opened. Current is injected at the C11 and C51 connections at the CT terminal block. Current flow is from C11 through the relay red-phase via connection C71 through the yellow-phase to C31 and to C70 via the short-circuit at the CT terminal block; down C70, through the short-circuit across the earth-fault element to C71 on the blue-phase element and through this element to C51. Note in practice the C71 connections may differ physically from that shown and this may change slightly the connections which need to be opened to connect the phase elements in series.

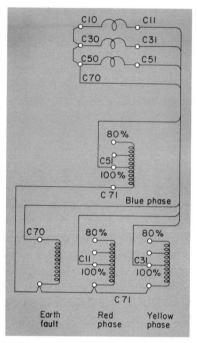

FIG. 12.8   MOTOR PROTECTION RELAY. TYPICAL CIRCUIT

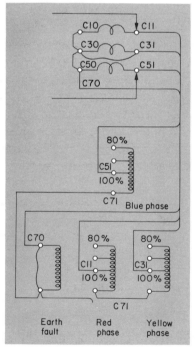

FIG. 12.9   MOTOR PROTECTION RELAY. MODIFICATIONS TO CIRCUIT
FOR TESTING

Procedure

Test setting				Timing			Inst. Relay Current		
Contact	Tap	Zero	100%	2×start	2×run	6×start	R	Y	E

The relay characteristic is shown on the nameplate for a particular overload setting, e.g. 115%. The test setting of the adjustable contact is this value. The overload elements should be on the 100% tap during the tests.

There is a square adjuster at the centre of each element and this is used to set the yellow-phase element so that zero is indicated on the % load scale and the contacts on the red- and blue-phase elements so that they lie symmetrically between the yellow element contacts.

With the relay cover replaced, twice relay normal current is applied and the time to operate is noted. The current is reduced to 100% and the position of the indicator on the % load scale and the relative positions of the contacts is examined. Adjustments are made to correct errors by moving the position of the heaters with respect to the bimetals. Starting with the yellow-phase element to adjust the indicator position to 100% and then the red- and blue-phase element to centralise the contact position. Alternatively the adjustment of the yellow-phase element could be made so that 100% corresponds to motor full-load current rather than relay 100% current. The indicator would then indicate the actual percentage load of the motor.

When adjustments are complete the current is increased to twice normal again and the time recorded in the "2X Run" space. Following this the relay is allowed to cool and then the 6× setting current time is recorded. During this test the contacts should be watched carefully to establish that the out-of-balance contacts do not touch.

The operating current of the instantaneous overcurrent elements and the earth-fault element are checked.

The relay settings are then applied and a check made to ascertain that the intended tap has been selected and then the nameplate disc is adjusted so that the correct tap is displayed.

Directional Relays

There is always difficulty in checking that directional relays are connected correctly and therefore the wiring should be comprehensively examined and circuits traced out. Following this, if possible,

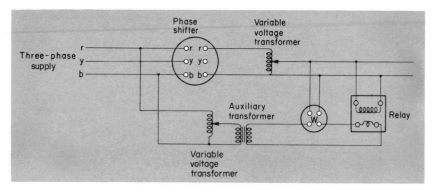

FIG. 12.10   TEST CIRCUIT USING A PHASE SHIFTER

tests should be made using a phase-shifter or other device to provide a phase shift and finally a check should be made under load conditions.

The use of a phase-shifter requires some explanation. Figure 12.10 shows the test circuit. The relay voltage coil is to be supplied through the phase-shifter and the current coil through transformers connected to the incoming supply. The variable voltage transformers and the transformer in the current circuit will introduce phase changes and therefore it is necessary to establish phase relationships at the relay. A wattmeter is used for this. The voltage coil is connected across the relay voltage coil and the current coil in series with the relay current coil. The voltage and current are adjusted to the nominal value and the phase angle changed in the lag direction until the wattmeter reads zero. The phase-shifter scale should be adjusted to 90° lagging. A check can be made that the wattmeter reads maximum at 0°, this is not easily determined, and that there is a zero reading at the 90° lead point.

Observing that the polarity is correct under load conditions requires the knowledge that the load is, without any doubt whatsoever, flowing in a certain direction. Even then confirmation may be difficult if the power flow is in the wrong direction for relay operation or if the phase angle of the load differs greatly from the optimum phase angle of the relay.

It should be borne in mind that the connections of a directional relay depend on the phase rotation of the system being correct. This has been known to be wrong.

Biased Differential Relays

To plot the bias slope characteristic of a biased differential relay

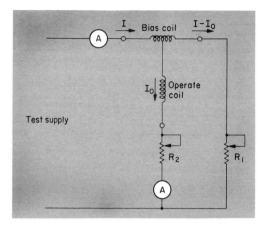

FIG. 12.11 BIASED DIFFERENTIAL RELAY TEST CIRCUIT

requires a circuit as shown in Fig. 12.11. The relative values of the resistors R1 and R2 are roughly the same ratio as the bias slope, e.g. for a bias slope of 20%

$$\frac{R1}{R2} = 0.2.$$

To determine the 1 A position on the characteristic. A current of, say, 1.1 A is injected and R2 adjusted until the relay operates. Let the operating current be I_0. Then the average bias current is

$$I_B = \frac{I + (I - I_0)}{2}$$

and the bias slope at I_B is I_0/I_B. The test is repeated for other values of I until the characteristic curve I_0 against I_B can be drawn.

General

On each test sheet there should also be space to enter the substation name, the circuit, the relay serial number and other relevant details. Spaces for results of insulation tests, CT magnetising characteristic tests and tripping tests should also be provided.

INSULATION TESTS

Insulation testing is performed by applying a d.c. voltage of, say, 1000 V between all circuits and earth and between each circuit and measuring the insulation resistance. This tests the insulation of the current transformers and wiring as well as the relay.

The general procedure is as follows:

Remove links and fuses in tripping and alarm circuits and to open the CT earthing links. Test and record the insulation resistance between all circuits and earth and between all circuits. It should be borne in mind that the tripping circuit and any alarm circuit which includes a current-breaker auxiliary switch will be in three parts. The positive, the negative and the circuit between the relay contact and the circuit-breaker auxiliary contact usually numbered K3.

Damage may be caused to static relays if they are subjected to insulation testing. Some relays can withstand the test voltage; some can if the test set does not generate spikes, i.e. the commutation spikes of a hand- or motor-driven generator, other relays require some modifications to the connections from the opening of a link in some cases to almost complete disconnection in others.

This means that reference to manufacturers' published data is essential to be certain of the steps to be taken to prevent damage.

TRIPPING CIRCUIT CHECK

All testing should culminate in trip circuits testing by tripping the circuit-breaker by operating the relay. Ideally the relay should not be operated by hand but sometimes this is unavoidable. When this is the case great care should be taken to ensure that the relay is not damaged.

ROUTINE MAINTENANCE TESTS

These tests can be as extensive or as limited as common sense dictates depending on the importance of the circuit. In general, however, they would be mainly confined to insulation testing of all the wiring, relay characteristic tests and tripping checks.

TEST EQUIPMENT

A test set to inject current into a protection system consists of a means of producing a variable current, an ammeter and a timer. The power requirements are about 3 kVA for primary injection and 1 kVA for secondary injection testing. In the latter case the power requirement can be reduced if only a limited range of relays are to be tested.

If a full range of 1 A and 5 A relays are to be tested the 1 A earth-fault relay set at 20% would require at 10× setting a current of 2 A and a voltage of

$$\tfrac{1}{2} \times \frac{3}{0.2} \times 10 = 75 \text{ v}$$

based on the relay having a 3 VA burden at setting and saturating to half the impedance at 10× setting.

At the other end of the scale a 5 A overcurrent relay on 200% tap would require a current of 100 A at a voltage of

$$\tfrac{1}{2} \times \frac{3}{10} \times 10 = 1.5 \text{ V.}$$

The VA is 150 in each case but to cater for the entire range would mean 50 − 1.5 V, 2 A windings which could be connected in series or parallel. In practice eight 10 V, 12.5 A windings would be used but this means a total rating of 1 kVA.

If only 1 A relays are to be tested then the requirement is a range of 75 V, 2 A to 7.5 V, 20 A. In this case four windings each of 20 V, 5 A could be used—a rating of 400 VA which means a test set of only half the size and weight. Similarly if only 5 A relays are to be tested four 4 V, 25 A windings could be used.

Figure 12.12 shows a basic test circuit. The current is varied by the variable voltage transformer and the resistor. The test transformer has a dual function, it provides the injection supply and also isolates the output from the mains so that there is no danger of the mains supply being short-circuited via the CT earthing circuit. Links are provided to allow series, parallel or series-parallel connections of the four windings to be made depending on the test current required.

Variable voltage transformers have a very poor wave-form when used at the low end of the range and as this may affect the performance of some relays, induction types in particular, a resistors placed in the circuit so that an increased voltage, with a consequent improved wave-form, should be used.

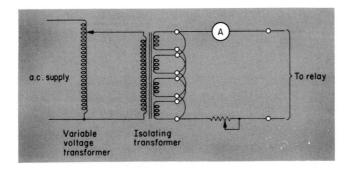

FIG. 12.12 TEST EQUIPMENT—BASIC DIAGRAM

It is useful to be able to connect directly to the output of the variable voltage transformer so that the test set can be used to provide the supply for CT magnetising characteristic checks. As this supply is not isolated from the mains it is important to know the polarity of the supply so that the line connection is not inadvertantly connected to earth. An indicator lamp connected between line and earth can be used to show correct polarity and that the test set is earthed.

A timing device is used which is arranged to start when the current is switched on and stop when the relay contacts close.

CARE OF PROTECTION RELAYS

Relays are generally of robust construction and not easily damaged after they have been installed. There is, however, some danger of damage before and during installation and during this time they should be treated like the measuring instruments that they are and should not be subjected to mechanical shock nor stored in unsuitable conditions. They should be handled with care and the cover should not be removed unless absolutely necessary. The removal of the cover during installation not only allows the ingress of dust, which is usually abundant during installation, but if drilling or filing is taking place nearby there is a danger that steel particles will enter the relay which sooner or later will be pulled into the air-gaps of electromagnet or permanent magnet and impair relay operation.

After installation there is not the same danger of damage but nevertheless relay covers should not be removed unnecessarily and any broken glasses should be replaced immediately. Before removing a cover the relay case should be thoroughly cleaned to remove all dust. Most damage to relays is caused by inexperience and therefore any work should be carried out by a skilled person.

Before any work on the relay is started the trip circuit should be isolated and a visual inspection made. Moving parts should be carefully cleaned with a feather and a piece of stiff card is useful in removing swarf from the magnet gaps. Relay bearings should not be lubricated as they are designed to be dry to eliminate the possibility of sticking after along period without movement.

Relay contacts should be inspected and only cleaned if there is discoloration to such an extent that may impair good contact. Cleaning should be by a soft paint brush dipped in clean trichlorethylene followed by wiping with a lint-free fabric or chamois leather and then burnished. If the contacts are slightly pitted they should be burnished smooth if possible. If not they should be replaced. Under no circumstances should abrasives be used as grit may become embedded in the contact face which will ultimately cause a failure.

Index

All or nothing relays   25
Arcing time-fuses   10
Attracted-armature relay   23
Auxiliary relays   26
Axial moving-coil relay   27

Balanced-beam relay   169
Balanced earth-fault protection   148
Balanced-voltage differential
    protection   154
Batteries   204
Biased differential protection   21,
    125, 224
Bimetal   6, 29
Buchholz protection   140
Busbar protection   115
    frame leakage   118
    fully discriminative   117
    mesh corner   124
    neutral check   119
    phase and earth fault scheme   120

Cables—fuses   13
    impedance   67
    sheath resistance   79
Capacitor voltage transformer   55
Care of relays   228
Check relays   167
Circulating current differential
    protection   154
Commissioning tests   212
Comparators   31
Contactors—protection by fuses   12
Contacts   33
Control circuits   200
Current transformers   40
    accuracy limit factor   46
    application   49
    burden   42

class X   47
    construction   40
    design   41
    e.m.f. equation   42
    equivalent circuit   43
    knee point flux   41
    knee point voltage   219
    magnetising characteristic   45
    magnetising current   43, 51
    magnetising curve   217
    open-circuit   45
    operation   43
    polarity check   212
    primary windings   49
    quadrature transformer   51
    ratio check   213
    residual connection   53, 113
    secondary rating   48
    short-time factor   46
    specification   47
    summation   51
    winding impedance   49

D'Arsonval moving-coil relay   27
Dashpot   7
Delta/star conversion   72
Differential protection   103
    feeder   153
    busbar   115
    generator   187
    motor   182
    transformer   130
Differential relay testing   224
Direct-acting trip   2
Directional relay testing   223
Discrimination   1
    mcb and fuse   3
    IDMT   87
Duplicate busbar protection   124

Earth—fault protection
  calculations 78
  feeders IDMT 98
  generators 189
  transformers 136, 146
Earthing 78
  generators 187
  transformers 79
Earthing resistor 79
Earth return impedance 80
Enclosures 34
Extremely inverse DMT relay 101

Failure of prime mover 195
Fault calculations 56
  cables 67
  current distribution 77
  delta-star conversion 74
  earth faults 78
  earth return impedance 80
  example 71
  generators 62
  motors 71
  sheath resistance 79
  source impedance 69
Feeder protection
  balanced voltage 154
  circulating current 154
  differential 153
  IDMT relays 87
  impedance 168
  intertripping 165
  pilot insulation level 157
  pilot wire characteristics 156
  Solkor 162
  summation transformer 154
  transformer-feeder 164
  Translay 160
  teed 164
Fuses 8
  arcing, pre-arcing and total time 10
  cables 13
  discrimination 11
  discrimination with IDMT relays 90
  I^2t values 9
  motors 12
  motor starting 8
  protection of contactors 12
  semi-conductors 13

Fusing current 8

Generator
  decrement curve 67
  earthing 186
  fault 62
  impedance 62
  sub-transient, transient and
    synchronous reactance 63
  time constants 64
  unbalanced loading 190
Generator protection 186
  differential 187
  earth fault 189
  failure of prime mover 195
  insulation failure 186
  loss of field 195
  negative phase sequence 190
  overcurrent 192
  overload 194
  overspeed 196
  overvoltage 197
  rotor earth fault 189
  stator protection 187
  typical scheme 198
  unbalanced loading 190
Generator-transformers 197
Generator unit transformer 199

High impedance relay 106
  application 109
  calculations 112
  knee point voltage 112
  residual connection 113
  setting 113
High-set relay 143

IDMT relay 81
  application 87
  calculation 92
  characteristic 82
  connections 82
  discrimination 84
  discrimination with fuses 90
  earth fault protection 98
  plug setting 81
  ring system 95
  settings 85

testing   220
time multiplier   83
time multiplier setting   86
Impedance
  cable   67
  feeder   62
  generator   62
  per cent   57
  per unit   58
  reactor   62
  source   69
  Sub-transient, transient and
    synchronous   63
  transformer   62
Impedance diagram   59
Impedance protection   168
Induced voltage in pilots   157
Induction motor   173
Induction cup relay   21
Induction relay   15
  application   18
  biased   21
  control   17
  operation   15
  single quantity measurement   20
Intertripping   165
Insulation tests   225
Intertripping   165

Knee point flux   41
Knee point voltage   219
KVAr relay   19

Loss of field   195
Loss of supply   183
Magnetising characteristic   217
Magnetising inrush   131
Maintenance   226
Measuring relay   26
Merz-Price protection   103, 125
Mesh-connected substation   124
Mho relay   172
Microprocessor-based relay   37
Miniature circuit-breaker   3
Motor fault contribution   71
Motor protection   173
  differential   182
  insulation failure   178
  loss of supply   183

overcurrent   178
overload   175
setting   179
testing   221
Moulded-case circuit-breaker   3
Moving-coil relay   26
MVA base   58

Negative phase sequence protection
  generators   190
  motors   177
Neutral check, busbar protection   119
Neutral earthing resistor   79
Non-linear resistor   107

Oil dashpot   7
Out-of-step relay   185
Overcurrent protection
  feeders   81
  generators   192
  transformers   143
Overcurrent relay   20
Overload devices   6
Overload protection
  generator   194
  motor   175
  transformer   144
Overvoltage protection generators
  197
Overvoltage relay   20

Phase-angle compensated relay   19
Pilot characteristics   156
Pilot supervision equipment   167
Plug setting, IDMT relay   81
Polarity check   213
Power factor correction   204
Power system diagram   59
Pre-arcing time—fuses   10
Primary injection testing   213
Product measurement   30
Protection of a typical substation   145

Ratio check   213
Rectifier-comparator bridge   31
Relay characteristic check   220
Relay design   32

Relays   15
   all-or-nothing   25
   attracted armature   23
   auxiliary   26
   axial moving-coil   27
   balanced beam   169
   buchholz   140
   check   167
   D'Arsonval moving-coil   27
   extremely inverse   101
   high impedance   106
   high-set   143
   IDMT   81
   induction cup   21
   induction disc   15
   KVAr   19
   measuring   26
   Mho   172
   micro-processor based   37
   moving-coil   26
   out-of-step   185
   overcurrent   20
   overvoltage   20
   phase-angle compensated   19
   standby earth fault   140, 151
   starting   167
   static   35
   thermal   29
   timing   31
   tripping   25
   undervoltage   20
   wattmetric   18
Residually connected high-impedance
   relay   113
Restricted earth-fault protection   146
Ring distribution system   95
Rotor earth-fault protection   189
Rough-balance protection   133

Semi-conductor protection   13
Sheath resistance, cable   79
Single quality measurement   30
Site tests   211
Slip-ring induction motor   175
Solkor protection   162
Source impedance   69
Starting relays   167
Static relays   35
Stator winding protection of a
   generator   187

Summation transformer
   differential protection   154
   power factor correction   207
Synchronous motor protection   184

Teed-feeder protection   164
Test equipment   226
Testing   210
   biased differential relay   224
   commissioning   212
   CT magnetising characteristic   217
   CT polarity check   212
   CT ratio check   214
   directional relays   223
   equipment   226
   insulation   225
   overcurrent relay   221
   primary injection   213
   Relay characteristic check   220
   secondary injection   220
   site tests   211
   thermal relay   221
   trip circuit check   226
   works tests   210
Thermal overload   6
Thermal relays   29
Thermal relay test   221
Thermal trip   2
Time-graded overcurrent protection
   81
Time-limit fuse   4
Time multiplier setting   86
Transformers
   faults   129
   impedance   62
   magnetising inrush   131
   tap-changing   132
Transformer-feeder protection   164
Transformer protection   129
   balanced earth fault   148
   biased differential   130
   buchholz   140
   differential   130
   directional overcurrent   149
   earth-fault   136
   harmonic bias   134
   high-set relay   143
   high-speed relay   134
   overcurrent   143, 149
   overload   144

restricted earth-fault 146
rough balance 133
standby earth-fault 140, 151
typical substation 145
Translay protection 161
Trip circuit supervision 202
Tripping circuits 200
Tripping circuit check 226
Tripping relay 25

Unbalanced loading
  generator 190
  motor 177
Unit protection 103

Unit transformers 199
Underpower relay 185
Undervoltage relay 20
Unsatisfactory operating conditions-
  generator 190

Very-inverse relay 100
Voltage transformer 52
  accuracy 53
  broken delta connection 53
  capacitor 55
  protection 53

Wattmetric relay 18